THE RACE GALLERY

Marek Kohn is a writer who studied neurobiology at Sussex University and whose work has appeared in the *Independent*, the *New Statesman*, *Arena* and many other publications. His widely praised history of the early British drug underground, *Dope Girls*, was published in 1992. His 'Technofile' column appears in the *Independent on Sunday*.

for Sue

Contents

Acknowledgments

Acknowledgments serve several purposes. They are a reader's guide to influences, a matter of etiquette, and an expression of genuine gratitude. In regard to the latter, I'd like to note that the responses of the vast majority of people I contacted (whether they are listed here or not) did much to help make writing this book a pleasurable as well as an intellectually rewarding experience. That has been an agreeable source of surprise, given the subject matter.

I would like to acknowledge the contributions made by particular individuals on particular themes: Chris Stringer on human evolution; Thomas Acton, Ivan Bernasovský, Moris Farhi, Ian Hancock, Milena Hübschmannová, Donald Kenrick, Sarabjit S. Mastana, Klára Orgovánová, Chris Powell and David Smith on the Roma; Georgia Dunston, Hank Greely, Fatimah Jackson and Alan Swedlund on the Human Genome Diversity Project; Ivan Čolović and Silva Mežnarić on nationalism in ex-Yugoslavia; Stephen Howe on Afrocentrism; Owen Anderson, Dave Hill and Kenan Malik on sport; and lastly, though they may not all wish to thank me, Pierre van den Berghe, Richard Lynn, J. Philippe Rushton and Vincent Sarich on their own work.

Thanks also to Martina Derviş for German and Anna Fedurcová for Slovak translation, to the Hypermedia Research Centre at the University of Westminster for World Wide Web facilities, and, for one thing and another, to John Archer, Rosalind Arden, John Bancroft, Jeanette Buirski at European Dialogue, Fulvio Caccia, Stephan Feuchtwang, Liz Fekete at Institute of Race Relations, Michael Goss, Ralph Holloway, Mike Hutchinson, Robert Kruszynski, Adam Kuper, Richard Newman, Polly Pattullo, Berndt Pickert at *die tageszeitung*, Martyn Rady, Paul Rambali, Gregory Rabess, Matt Seaton, David Skinner, Michael Shermer, Johann Szilvássy, Boyd Tonkin, Rachel Tritt, Ron Walters, Nick White, John Witton, Elizabeth Young, the staff of the Museum of Mankind Library, Die Grüne Alternative in Vienna, *Profil*, Rural Advancement Foundation International, *Searchlight*, Survival International and the World Council of Indigenous Peoples.

Special thanks to several scientists who were particularly generous with their time: Chris Stringer, Steve Jones, Ivan Bernasovský, Shomarka Keita and Fatimah Jackson; also to Stephen Howe and Kenan Malik for letting me read their own (excellent) manuscripts, and to my editor, Neil Belton, for his role in developing this project and his commitment to it.

Above all, thanks to my wife, Sue Matthias Kohn; for everything.

Introduction

THIS BOOK'S PRIMARY subject is a particular way of thinking and talking about race: the scientific way. But *The Race Gallery*'s broader theme is that of how we think and talk about race in general. We do so at inordinate length, yet with extreme anxiety: we feel that the subject is covered by a taboo, but we don't know exactly what the rules of the taboo are. It seems important, if not obligatory, to discuss cultural differences, but dangerous even to mention physical differences. At the same time that we feel compelled to do it, we are profoundly uncertain about how to think about race.

Science's contribution to resolving our unease has been the message that, in scientific terms, race is of minimal importance, if not a delusion altogether. The construction and development of this message has been one of science's most significant contributions to society in the half-century that has elapsed since the end of the Second World War, though the fact has not been generally appreciated. Over the last twenty years, the line has hardened: the concept of race has largely vanished from textbooks, except to be labelled obsolete. A century ago, equivalent books would have spoken of little else.

If race is understood to be a scientific solecism, it is natural for lay people to infer that there can be no meaningful differences between different varieties of humankind. But the rejection of the race concept refers to ideas about classification, not variation. The distinction was visible – if unremarked – in the reaction to Richard J. Herrnstein and Charles Murray's *The Bell Curve*, which was published in 1994. Herrnstein and Murray presented a body of evidence, based on IQ scores, which many psychologists believe indicates genetic differences in mental ability between blacks and whites. While some commentators made trenchant criticisms of the internal logic of the argument, there was also a flurry of articles with titles like 'The Fallacy of Race'. These generally emphasised that the varieties of humankind shade into each other, and that the differences within groups tend to be much greater than those between them. But the hereditarian psychologists would not disagree with that at all. They accept that the dividing line between black and white is artificial, but argue that once such a line is drawn, inequalities are revealed. When you stop to look, the comfort that science offers anti-racism is far from complete. Race is a fallacy, but human uniformity is a *non sequitur*.

Science has not wrapped up the issue of race and consigned it to the dustbin of history. Rather, science has refuted certain aspects of race as it was conceived in the nineteenth century, and used the refutation as the basis for a moral and political intervention on the side of anti-racism. While the traditional race concept, that humanity could be divided into clearly demarcated varieties, which possibly did not even share a common origin, underwent a long decline because of its internal contradictions, it was not decisively rejected until the defeat of Nazism. In other words, the eclipse of the race concept was the result of history as well as logic.

The point of departure for *The Race Gallery* is the fact that scientific anti-racism is an element of the post-war order which also included the Soviet bloc and Western European social democracy, and the possibility that it may also turn out to be temporary. *The Race Gallery* explores the presence of race in scientific discourse, and the presence of scientific (or pseudoscientific) themes in political rhetoric and popular ideas of race. Its purpose is twofold: to explore how race has persisted in science despite its repression, and to consider its future prospects.

One of the book's basic premises is that science is embedded in society, and has to be understood as such, but also needs to be considered in its own terms. Historians have argued convincingly that scientific racism arose under particular historical conditions, reflecting the worldview of Northern Europeans who enjoyed political and economic dominance over the darker-skinned peoples of the world. It was self-serving, self-centred, and used to justify great cruelty and oppression – but that does not necessarily mean it was wrong. If a scientific argument is shown to be rooted in a racist tradition, it should be regarded as dubious, but it cannot be deemed disproven. Refutation can only take place according to science's internal procedures. That, incidentally, is why hereditarian scientists are so irked by Stephen Jay Gould's book *The Mismeasure of Man*: it examined both the historical context of race science, and the data too.

Another of *The Race Gallery*'s main premises is that science actually has many voices, not one: the voice that speaks to the world through Unesco or the Discovery Channel, affirming the biological meaninglessness of race, may be less important in a particular place or zone of influence than other voices with a different message. *The Race Gallery* opens with an illustration of how traditional race science, partially updated, has managed to survive in one particular European institution. It goes on to give an account of how scientific anti-racism was established, and how a minority current resisted the new order.

The book then looks at areas in which race remains both embedded and, to a greater or lesser extent, acknowledged as a matter of controversy. These include palaeoanthropology (the study of fossil humanity) and physiology, with particular reference to the question of whether various human groups have natural advantages in certain forms of sport. Both fields contain arguments implying that the details of evolution could have left different human groups with significantly different capacities; which leads on to the issue of mental abilities.

A discussion of the implications of IQ testing – which, especially since *The Bell Curve*, is wrongly assumed to be more or less the whole of the race question in science – then leads to an examination of the work of psychologists who have linked IQ and evolutionary theory to reinvent nineteenth-century scientific concepts of racial hierarchy. This is followed by a look at Afrocentrism, the mirror image of white race science. Finally, Part One closes with a brief consideration of racial dimensions in evolutionary accounts of human behaviour. An important theme in human sociobiology is the idea that, although

...ups may not be significantly different, they have an innate
...o believe that they are: in other words, ethnocentrism is
...our biological nature.

r.. ...ne is about history and theory. Part Two is about practice
and diversity: the first two chapters present a case study of how race
remains embedded in scientific practice, but is not acknowledged.
Part One deals with recognised problems, whereas much of Part
Two is about unexamined assumptions.

The subject of the case study is the Romani people, or Gypsies. I
found my way to them by reasoning that the place to look for actually
existing race science is among the politically weakest of peoples, and I
duly found plenty of evidence in support of the hypothesis. (I also
came to perceive something of the extraordinary depth of hostility
that Roma face wherever they live, and to fear that their unhappy
history in Europe may prefigure the near future for people of
different cultures who have come to the continent in more recent
times.)

The first chapter on the Roma explores how medical science is
influenced by attitudes held in the dominant society as a whole; in
turn, it influences popular ideas and political rhetoric, forming a
circuit of exchange. Biologistic jargon has been used against the
Roma by demagogic politicians in Eastern Europe, and there is
evidence that they have been subjected to a campaign of racial
eugenics.

The Roma also attract scientific attention as a test-bed for the
study of 'ethnogenesis' (the origins of ethnic groups) by the
examination of their genetic characteristics. These studies, the
subject of the second chapter, invariably support what historical and
linguistic study shows: that the Roma came from India. As with the
medical findings, such information can be interpreted in different
ways. For Roma, it may strengthen their sense of identity; for racists
– including right-wing terrorists – it suggests a place to which they
can demand the Roma return, seven hundred years later.

Although the focus of the Romani chapters is in Slovakia, one of
the studies discussed took place in Wales. It led to a broader project
on Welsh population genetics, with a surprisingly traditional
approach. Chapter 11 discusses this work in relation to other visions

of the links between people and territories, including the use of biological rhetoric by contemporary Serbian nationalists. Genetic diversity studies promise a new synthesis of genes, history, linguistics and archaeology. They offer the possibility of demystifying the origins of ethnic groups, but they can also be used to perpetuate traditional Romantic notions of ethnic purity.

The following chapter describes the furious controversy surrounding the Human Genome Diversity Project, which plans to gather DNA samples from indigenous peoples around the world. It goes on to look at the idea that the history of a people may be read in its genes; specifically, that slavery has affected the genetic characteristics of African-Americans. The possibility of a positive, non-racist appreciation of human biological diversity then informs – though sadly does not dominate – the conclusion of *The Race Gallery*.

Most books would need no further introduction, but if that were the case here, this book need not have been written. In some readers, the very fact of its subject matter will raise suspicions about its intentions. I have little to say about other people's motivation in the following pages. A scientist's feelings about people of other races may be relevant to his scientific work, but the question is apt to turn into a red herring. Similar arguments apply to the question of motive as to that of historical context. Scientific texts cannot be satisfactorily analysed in terms of the supposed motives of their authors. If it were established that a scientist had hidden Dr Mengele in his garden shed – and indeed, rumours along similar lines have circulated about one of the individuals mentioned in this book – that would not refute any of the scientist's hypotheses, or cast any doubt on his data. To cast moral aspersions, by calling somebody a racist, is not the same as proving them wrong.

I think, however, that it is as well for me to say something about my own motivation. This is partly because I do not want people to infer that since this book deals with race in a manner which deviates from the prevailing anti-racist style, it must have been written by a racist. A brief note of my own perspectives, or biases, may also serve to set the book's arguments in context. They are those of a European in whose childhood the post-war order seemed permanent: both the good, as manifested in the Welfare State, and the bad, as symbolised by the Iron Curtain. Part of that order, the result of war and technological progress, was the ideal of human unity. People have perceived its expression in blue helmets, in the image of the planet photographed from an Apollo spacecraft, or in a globally broadcast charity concert. It remains elusive, but I still believe in its essential truth, and in the corollary that racism is one of the most evil forces abroad in the world.

It follows that I value the anti-racist line adopted by the dominant currents of science after the war. But scientific anti-racism is a social construct, just like scientific racism, and like most social constructs, it needs to renew itself as society develops. That is not possible if the scientific dimensions of race cannot be discussed frankly. The social developments of recent years include the onward marches of genetic determinism and evolutionary accounts of human behaviour, the explosive growth of ethnocentric nationalism, and an isolationist mood among affluent whites, who may well be inclined to rationalise their reluctance to share resources and space with the poor by adopting scientific claims of innate black (and poor white) inferiority. In the face of these currents, denial is the kind of response that gets wet liberalism a bad name.

There is also a principle involved here; the one embodied in the First Amendment to the United States Constitution. Race scientists should enjoy freedom of expression – and these scientists should be prepared to participate in the free circulation of ideas upon which intellectual life depends. In this regard, I would like to acknowledge the generous manner in which Professors Richard Lynn and J. Philippe Rushton, whose work receives extensive and highly critical attention in this book, continued to provide reprints and information after they realised I was on the wrong side. Their attitude stands in marked contrast to that of their colleague Dr Roger Pearson, who gave me to understand that he was unwilling to enter into discussion with somebody who held an ethical objection to racism.

I should also state that I am not a scientist, but a writer with a science degree, which I believe places me in a usefully intermediate position between the culture of science and a lay public. When scientists write popular books, the relationship between them and their readers is that of teacher and pupils. Their natural role is to make authoritative pronouncements. Many scientists have pronounced race science invalid, however, and yet it persists. As an outsider with no pretensions to scientific authority, I am more concerned with the balance of forces which govern the power of race science.

This book is intended to explore those forces, and to provide an intelligent commentary on scientific ideas, pointing out their important features and illustrating objections to them, but it is not what scientists call a review of the literature. Nor is it a popular scientific textbook, or a review of the history of race science: there are already many good books addressing those needs. In discussing how we think about certain aspects of race, it uses both scientific and non-scientific techniques. Just as human biological diversity varies continuously, so does culture; the barriers between science and non-science are as artificial as those posited between human groups.

When Steve Jones, Professor of Genetics at the Galton Laboratory of University College London, remarked confidently to me that race would not return to science, I objected that it was already thriving in psychology. Yes, he replied, but not in science. I had a good deal of sympathy for this viewpoint when I began working on this book. Now that it is finished, I have a good deal more. However, I have tried to keep it to myself. A common critical tactic, especially among scientists, is to declare that the object of criticism is 'not science'. This denies legitimacy to the work in question, and it also gets science off the hook. I have tried to show that, rather than being merely the enthusiasm of a few maverick scholars, race is embedded in many different scientific fields. In general, I have chosen to regard any kind of 'ologist' as some sort of scientist, and to treat as scientific any work that goes through the motions of scientific protocol. If somebody enjoys the institutional and cultural trappings of a scientist, then he or she is a scientist.

Even after discarding intellectual criteria in favour of a sociological definition, however, there are limits. When extreme Afrocentrists talk about the paranormal properties of the skin pigment melanin, there seems little point in refraining from calling them pseudoscientific. And although there is a grey area between science and the rest of culture, I can't envisage a definition of science broad enough to embrace the 21-point list of criteria of civilisation cited by Professors Lynn and Rushton, even though it is integral to their theories.

As for the term 'race', I prefer not to limit its range of meanings: I hope that where it occurs in this book, the sense in which it is meant will be clear from the context. What it does not mean, however, is that I take race to be a biologically meaningful concept. One of the most important messages of this book is that a revival of racial science remains possible despite the rejection of traditional scientific concepts of race. The old racial categories were just the suitcases, not the whole of the baggage. What mattered were the contents, and these may find new niches within any of the scientific fields associated with the division of humankind into groups.

Race, moreover, is a peculiarly forceful word. Unlike expressions such as 'ethnic group', appropriate as they are in many contexts, it reminds us of the subject's scale; of its enormity. It challenges us to face the matter.

PART ONE

I

Rassensaal

ALL SMILES; THE peoples of the world are arranged here according to their differences, but they are assembled in friendship, or at least cordiality. Their radiance requires that we like them, or at least that we contemplate them in a spirit of benign generality; that we can encounter each other in all our variety, and pass on unscathed.

Whatever the original feelings behind all these smiles, they are concentrated here to serve the single purpose of outshining the grins of a hundred, perhaps two hundred, human skulls. In this parade of humankind, the photographs mediate between the living viewers and the reminders of what they will become. They also mediate between science and people, offering reassurance that science too has a human face, even though its real concern is with the metrics and characters of the skull underneath.

The proportions of the square outside, expansive enough for a regiment, also suggest a parade. Indeed, this entire district of the city is an imperial metropolis, monumental bastions separated by parade grounds. A million people are said to have filled one once, when Adolf Hitler addressed them in the Heldenplatz, but they did not embody the human variety on display in this hall. Rather the opposite, in keeping with their *raison d'être*.

The hall's proportions are imperial too. Around the top of the walls is a frieze, depicting in tableaux the variety of peoples outside Europe. When painted, these represented Europe's reach and the extent of its dominance. Human varieties were vertically arranged,

9

with the northern European at the top. Now the specimen 'Nordid' grins, unpersuasively, presenting himself in the collar, tie and blow-dried hair of a salesman, not the locks and scowl of a warrior. The smiles are tokens of equality; between the Nordid, the Alpinid, the Bushman, the Gypsy, the Kafrid.

None the less, this gallery of human diversity insists above all that differences are real, and that humans can be arranged in order according to them. The use of single examples for each variety, such as the blond Nordic male, expresses faith in the notion of racial type. The essence of type resides in the skull; its length, its breadth, the size of the teeth, the height of the brow, the prominence of the chin, and Latinate details of its structure, from bregma to foramen magnum, apparent only to the anatomist.

Traditional typology, in which human groups are divided into 'types' according to physical traits, proposes a connection between these cranial nuances and the characteristics of the type as a whole. In one corner of the hall, a table lists the 'important morphological and physiological characteristics of the great races', from 'cranial index' and 'hair colour' to 'pathological dispositions'. But no further. Out of the tableaux up above, and the showcases of skulls all around, the silent invitation fills the hall: Dear visitor, we invite you to read between the lines.

A race gallery seems an improbable fixture for a major museum a few years short of the close of the twentieth century. When they exhibit national collections of global significance, metropolitan museums embody ideals. Once upon a time, seeking to impose order upon human variety seemed an honourable project; today, seeking to divide humankind into groups appears retrograde.

It also appears superfluous, since the world now offers so much evidence to suggest that fragmentation is the natural state of chaos to which humans revert if unchecked. Alternatively, it may appear sinister, given the abundance of evidence for the case that ethnic division is an unnatural condition into which people are manipulated by malign elements.

Against these tendencies stands civilisation, which incorporates knowledge, enlightenment, science. Its task, surely, is to impart the message that humanity is essentially one; that at the fundamental level of biology, differences between human groups are too trivial to dwell upon.

Since the end of the Second World War, the dominant voices of science have provided just such a guarantee. The basis of this assurance is that races themselves are illusory. In point of biological fact, the textbooks affirm, there are no clear lines between human populations; all merge into each other. Humanity is now conceived not as a range of types, but as a network of interconnected breeding populations. Instead of separating out the types within a population, the population is considered as a single entity within which individuals exchange genes; the range of characteristics previously grouped into types is regarded as the range of variation within the population as a whole. Rather than conceiving the British population, for example, as a mixture of types, or races, it is understood as a dynamic system containing a range of variation that extends in eye colour from brown to blue, in hair colour from black to yellow, and so on.

As well as proving scientifically useful, as a way of understanding population genetics, the 'populational' concept is better suited to a world of rapid migration and interbreeding. In this perspective, people of African or East Asian descent represent an extension of the range of variation found in the British population, rather than an intrusion of alien races. Admittedly, for practical purposes, it is often useful to divide humankind into Caucasoids, Mongoloids and Negroids, but no further. The detailed ordering of humanity belongs to the past, which means to a current in history that culminated in, and ended with, Hitler's regime.

None the less, half a century after the death of that Austrian, a race exhibit remains open to the public in Vienna's Natural History Museum. On the museum plan, Hall XVII is labelled *Rassen des Menschen*; Races of Man; to museum staff, it is known as the *Rassensaal*, the Race Hall. In British museum terminology, it would be called the Race Gallery.

It imposes a hierarchical order on the 'human family', though only in the formal sense that it uses a system in which the species is divided first into the 'great races' – Europid, Mongolid, Negrid – and then into types within the grand divisions. Among the Europids, for example, may be found Alpinids, Dinarids and Mediterranids. This classification system derives originally from schemes set out in the late nineteenth century, notably by the American William Z. Ripley, who divided the peoples of Europe into Teutons, Alpines and Mediterraneans, and the Russo-French Joseph Deniker, who identified ten European races.[1] All three of the main European races – Nordics (Ripley's Teutons), Alpines and Mediterraneans – were understood to be found in Britain: the Nordics were believed to be concentrated in the English ruling class, while Wales was filled with small dark Mediterraneans.

Perhaps the last major exposition of this system was the American anthropologist Carleton Coon's study *The Races of Europe*, which appeared on the eve of the Second World War.[2] Considering the Balkans, he observed that 'the Bosnians serve racially as an approach to the nucleus of Dinaric gigantism in Montenegro'. Coon also stated that Bosnian Catholics – that is, Croats – had broader skulls than Bosnian Muslims. The Catholics had a mean cephalic index, a measure of the ratio of skull breadth to length, of 86; that of the Muslims was 84. The Catholics were taller and lighter-skinned than the other ethnic groups of Bosnia; Coon considered them purer descendants of the region's original population, showing fewer signs of 'outside' Slavic and Turkish influences.

The fascination of origins, and of characters which seem to embody the essence of a people, can be detected next door to the *Rassensaal*, in the gallery devoted to human evolution. After dealing with the evolution of the human species, it moves on to ethnogenesis. This is the process of the emergence of ethnic groups, or what some would call the birth of nations. 'In the fourth century the GERMANS arose,' says a label in one of the showcases. 'Of their physical characteristics already the ancient writers (Tacitus) mentioned the tall stature, the blond hair and the blue eyes.' Actually, Tacitus said they had red hair, but in other respects his original delineation of the Nordic stereotype ('untainted by marriage with others, a peculiar people and pure') has endured.[3]

As in its neighbour, the building bricks of the human evolution gallery are skulls. Although there must be several hundred on display in the two rooms, these represent only a tiny percentage of the collection. In the part of the evolution gallery devoted to explaining the principles of physical anthropology, it is noted that the Anthropology Department possesses around 25,000 skeletons. The persistence of the two galleries is an assertion that traditional physical anthropology, founded upon the comparative analysis of skeletons, retains its validity today.

That assertion is a statement of defiance, in the face of competing disciplines and political opposition. The survival of the Race Gallery of Vienna into the middle of the 1990s depended upon the interplay of a number of different forces.

One of these was undoubtedly inertia. By the mid-1990s, the Natural History Museum of Vienna looks about twenty years overdue for the refurbishment which is at last under way. An attempt at a human biology display in the contemporary museum idiom, appealing to a child's sense of spectacle rather than to nineteenth-century scholarly aesthetics, succeeds only in emphasising the museum's exhaustion. Elsewhere, stuffed animals are crammed hide to horn in the showcases of a dingy gallery, like a trainload of livestock forgotten and left to mummify in a siding.

Across the square, the Art History Museum is just as crammed, but with masterpieces, and it brims with the self-confidence that comes with being one of Europe's great art museums. Nearby, the modernised sections of the Ethnographic Museum are harmoniously integrated with the original interior. It is not simply that the Ethnographic Museum is a finer building than the Natural History Museum, or that it has employed better designers in recent years, but also that it has been readier to grasp new ways of representing humankind. There is nothing in its exhibits that deviates significantly from the norms that prevail in the cultural anthropology of the English-speaking world.

The persistence of the Race Gallery, by contrast, was the result not just of tardiness in modernisation, but of resistance to newer concepts of human diversity. It roused the ire of anti-racist students in the 1970s, but subsequently escaped external pressure until 1993. In March of that year, the magazine *Profil* ran a strongly worded feature which invoked the spectre of Nazi race science, under the headline

'Like in Adolf's Time'.[4] Then in August, an article appeared in the newspaper *Kurier* which drew upon the authority of Johann Szilvássy, the Museum's Curator of Anthropology. Its theme was that Austria's population was more ethnically mixed than ever before, and quoted remarks by Szilvássy about the prevalence among people from south-eastern Europe of inbreeding, which 'leads to the appearance of recessive genetic characteristics'. *Profil* returned to the attack, joined by Green deputies in the Vienna city hall and the national parliament, who tabled questions about the *Rassensaal* and the *Kurier* article.[5]

In the same month, the controversy passed beyond the border, when a letter from the British anthropologist Adam Kuper appeared in the journal *Nature*, calling for an international campaign to ask the Museum to revise its exhibit.[6] Kuper included the human evolution gallery in his criticism, singling out a section of the display which juxtaposed the skulls of a chimpanzee, an australopithecine – the earliest member of the evolutionary branch leading to modern humans – and a Southern African Bushman. The implication, taken for granted in the bygone era of race science, was that the Bushmen (now more often known to outsiders as the San) were more primitive in form than other living human groups.

Szilvássy continued to defend the exhibit. 'I am aware of criticism from Left-oriented groups who accuse us of Nazi links,' he told the London *Sunday Telegraph*. 'But these people are quite wrong. All we are trying to do is represent systematically each type of human being.'[7] The triad of skulls, implying a racial hierarchy rather than a range of equal varieties, undermined this contention. It was subsequently removed from display.

The balance of forces turned against the race gallery in November, when Vice Chancellor Erhard Busek, in his secondary capacity as Minister of Science and Research, replied to the parliamentary Greens' queries. He announced that he had ordered a new anthropological exhibit.[8] In the meantime, the existing galleries remain open; the revision is planned to take place in the next few years, but no date has been set.[9] An unobtrusive panel in the corner of the Race Gallery noted that the mission of the Anthropology Department has been to describe variations in human morphology, or form, and to explain their causes. 'Only through consciously recognising and understanding differences in characteristics are we able truly to live in a spirit of tolerance towards strangers,' it asserts.

This is too striking a claim to pass unremarked, though the text of the panel itself offers no further elaboration. One might quibble with the assumption that the world divides into 'us', encompassing the authors and the readers of the text, and 'foreigners'. And the idea of 'tolerance', even if intended to mean nothing more than harmony, still hints that there is something about the object of toleration that tries 'our' patience.

But the important point is the substance of the statement: that the desired state of relations between ethnic groups depends on the understanding of differences in characteristics. In a different context, this would be a bland expression of the multiculturalist proposition that once people learn about each other's ways, they will get along. A very different claim is being made here: that the differences between human populations which need to be understood are physical ones. But why? If the gallery is endorsing the anti-racist scientific position that we are all one – including foreigners – then it is hard to see the point of learning about the minor variations between groups.

On the other hand, a classification system, or taxonomy, of human diversity makes a great deal of sense if the underlying assumption is that variations in physical detail are linked to variations in behaviour. Then there might well be a need for 'tolerance' between races, and an understanding of differences in characteristics would certainly be needed to achieve it.

The panel goes on to say that the curatorial staff recognise the need for the exhibit to reflect recent advances in human molecular biology; in other words, to convey a sense of human genetic diversity. The gallery would therefore be updated. Whether the genetic display would replace the morphological one is not stated, but if an understanding of differences in characteristics is indeed vital, logic would seem to demand that the skulls remain.

For the Department, the notice concludes, it is a historical imperative that the updated exhibit should describe how a 'value-neutral biological race-concept was turned into racism and abused to legitimate the extermination and expulsion of entire peoples under the guise of science'. Despite its rejection of racism, this statement reveals that the Department remains committed to a highly conservative position. The dominant view today, in science as a whole and the opinion received from it, is that races either do not exist or that they are of very limited use as a scientific concept. That view has been influenced by historians of science, who have argued that, rather than hijacking it at a later stage, racist ideologies produced race science in the first place.

The sense in which a biological race-concept can be value-neutral (*wertneutral*) is in any case restricted. If the race-concept includes the idea that human groups may differ in their mental capacities or in their psychological natures, as it traditionally did, then it undoubtedly has social and moral implications. The most that can be claimed is that its practitioners were and are unaffected, consciously or unconsciously, by any considerations other than those of objective scientific truth; except possibly during the period when the concept contributed to the process of genocide.

A more modest claim, advanced by a number of scientists in the post-war period, is that while race science was indeed rooted in racism, it could be purged of racist values to produce a neutral, objective systematics of human biological diversity. This line offers the possibility of reconciliation with prevailing views, but the Natural History Museum statement indicates a more profound commitment to the biological race concept. It implies that race science was tainted only by corruption, not original sin.

The Race Gallery was not created in isolation. Sympathy for traditional race science persists in the Anthropological Society of Vienna; Eike-Meinrad Winkler of the University of Vienna Institute of Human Biology argues that typology was never really at odds with the populational concepts that have largely superseded it, and that race biology, pursued with 'cool curiosity', can be separated from racism.[10] Such perspectives are not confined to this one Central European city, either. It is a continuing current in German-speaking anthropology, which effectively extends not just across Germany and Austria, but across much of the former Communist countries of East-Central Europe. Despite the Cold War, German and Austrian traditions exerted a major intellectual influence on East European scientists, just as Germany and Austria now exert a major economic influence on the region. In science, *Mitteleuropa* never really vanished.

Thus, in a scientific pamphlet from the Eastern Slovak city of Košice, published in 1994, the section on the 'anthropological characteristics of Roms (Gypsies)' says that 'as for the typological Roms composition, Indo-Afghanistan and Iran-Afghanistan admixtures are especially characteristic beside the mediteranoid, oriental, sublapoid, veddoidal and other compenents'. The first of three supporting references is Deniker's work *Les races et les peuples de la Terre*, published in 1926.[11]

Deniker's continuing influence is also acknowledged in a paper, *Morphological Properties of Yugoslav Population Groups*, from what in 1990 was the Socialist Republic of Serbia: 'His classification is still in use,' observes the author, Božina M. Ivanović, 'since the problem of human races [sic] classification is not as yet resolved, despite considerable advancement of anthropological science.' And despite the observation on the following page that 'the traditional typological conception' is 'outgrown today'.[12]

Superannuation does not appear to have diminished the baroque variety of typological schemes. Reviewing studies of ancient remains, in a paper published in 1988 as part of a project called 'Ethnoanthropological Characteristics of the Serbian Population', Ivanović's colleague Živko Mikić identified no less than twenty types, including Dinaric, Mediterranean-Atlantomediterranean, Mediterranean-Nordic, Mediterranean-Cromagnon, Alpine, Protonordic and Nordic, Atlantic-Nordic, Nordic-Cromagnon, robust leptodolichomorphic, Eastern European, palaeo-European (and Siberian), Baltic, Mongolian, Turanian and Pamirian.[13]

Ivanović follows Carleton Coon in noting that 'head length varies from one Yugoslav nation to another', and that 'Yugoslavs differ even more in regard to the head breadth'. Overall, Croats are characterised by 'tall stature and body mass', while in Serbia, 'the males are of medium and tall stature, and of strong physical constitution'. In Bosnia and Herzegovina, 'organized investigation of contemporary population groups' morphological properties have only recently begun'.

Within a couple of years, the findings of these scientific initiatives would be augmented by a folk anthropology which found its most fertile soil on the battlefields. Physical stereotypes of the enemy were particularly prevalent among soldiers: Croats envisaged their Serbian enemy as tall and blond; Serbs saw the Croats as tall and dark. Like Carleton Coon, the Serbs used darkness to signify that the Croats were less Slavic than themselves. In their own way, they were creating a front-line race gallery.

The persistence of Ripleyan and Denikerian typology illustrates a fundamental point: that science is not a monolithic entity with a single voice and a single centre of power. It just looks that way. The illusion is persuasive because so much work goes into maintaining it. Scientific institutions and the media share an interest in presenting scientific judgments as objective, and therefore singular. It is more satisfactory all round to report that 'scientists believe' X, or 'modern science has proved' Y, than to try to represent complex arguments on subjects that are hard enough for lay people to understand as it is.

Among social scientists and those suspicious of science, there is a widespread understanding that science is not simply objective knowledge, detached from society as a whole. Scientific truth is seen as relative rather than absolute, and biased towards the culture that produces it. Prophets of the New Age anticipate the imminent replacement of orthodox science by a new 'paradigm' of one sort or another; possibly humanistic, holistic, metaphysical, non-Western, or all and more of these. Scientific pronouncements are subjected to close scrutiny, and if doubt remains, they are not given the benefit of it.

There is one exception to this general rule, however. Without a moment's reflection, social scientists will invoke the authority of science to assert that racial science is 'spurious' or 'disproved'. This, of course, supports what they wish to believe. And as the next chapter will show, it is also the result of a highly successful project, led by scientists working through Unesco, to build a post-war anti-racist consensus upon a scientific foundation.

The anti-racist scientific perspective is now a pillar of the ideals professed by the 'international community'. But there is more to the world than globalistic rhetoric, the media and the social science establishment. Just as the United Nations may speak with one voice, but be unable to impose its will on small armed bands, so local centres of scientific power may be more important in their area of influence than the voice of the international consensus. Scientists in Slovakia affect the lives of Slovak Roma far more than scientists in New York or any other major international centre. It is Slovak scientists who make recommendations about Romani health care and education, and whose findings may seep into local media and political discourse.

The local area of influence may not be territorial, but that of a discipline. A survey of American psychologists published in the late 1980s found that about half of them believed that the differences in IQ scores observed between different races were partly genetic in origin, while only 15 per cent considered them to be entirely caused by environmental factors.[14] It is questionable whether such a view would prevail at a meeting of the American Association for the Advancement of Sciences, or even in the street outside. What matters much more is that psychologists have a strong influence upon educational practices. They also enjoy effective lines of communication to the policy intelligentsia, and thence to the media. The greater the isolation of psychologists from other relevant disciplines, such as genetics, the greater the part race concepts are likely to play in their thinking.

Psychologists tend to base their racial classifications on categories defined socially, rather than by physical or biological anthropology. Yet their findings have evidently persuaded many of them that races are both real and intellectually unequal. A handful of them have gone much further. These scientists have devised an evolutionary scenario which seeks to explain not just intellectual differences, but behavioural traits such as aggression and sexual promiscuity, according to the pressures of natural selection that they believe shaped the great races.

The starting point for this school of thought is not greatly at odds with mainstream thought, since its race concept is based on the three familiar divisions of Caucasoid, Mongoloid and Negroid. The difference is that while the mainstream acknowledges these divisions only diffidently, or not at all, the new race scientists claim that biological differences between these groups are responsible for the differences in their historical fortunes.

Although this school is proud of its roots in the racial scientific tradition, its models illustrate that the persistence of race science does not depend on the survival of pre-war theoretical structures. Deniker's classification system may still be used in Central Europe, but it is by no means the most important means of sorting humans into groups. Even where it appears in contemporary scientific papers, it may only serve as an adjunct to other schemes; as in the Slovak pamphlet mentioned above, which is mainly concerned with comparisons of blood groups among different Romani communities.

In such literature, quoting Deniker serves to indicate the broad tradition that informs the work. But new race galleries can be built without it. Among the tools adapted by the new race scientists are a theory of reproductive dynamics dating from the 1960s, and the 'African Eve' model of modern human origins, which emerged from molecular biology in the 1980s. The 'Eve' hypothesis, which proposes that all living people are descended from a single woman who lived in Africa, is usually seen to affirm human unity. As part of a wider set of ideas about modern humans' African origins, it contradicts the rival 'multiregional' model, which is based on the idea of race. In adopting Eve for their own purposes, the neoracial scientists show that the association between anti-racism and modern scientific thinking cannot be taken for granted. They eagerly anticipate that the Human Genome Project will add grist to their mill.

Mainstream geneticists ridicule such claims. The only possible effect of the data from the Human Genome Project, they maintain, will be to confirm the impossibility of dividing humanity into biologically meaningful subgroups. The Project will fulfil its international duty by affirming the essential oneness of humankind.

The goal of the Human Genome Project is to draw up a specimen of the entire DNA code for a human being. It will not be derived from an actual individual, however, but from thousands of samples. Most of these will be from people of European descent, reflecting the populations of the areas in which most of the research will be conducted. In order that the project might include the dimension of human genetic diversity, it was proposed that systematic regional sampling be undertaken, as part of an allied Human Genome Diversity Project. It was also pointed out that certain populations were unlikely to be available for sampling much longer. Theoretical and practical considerations converged, with the result that instead of aiming to cover the whole inhabited world, the HGD Project decided to concentrate on indigenous peoples.

The intended subjects of the research failed to see the project as a benign exercise in demonstrating human unity. Instead, they regard it as one more site of struggle. They are already involved in struggles over rights to land and mineral resources. They are gearing up to struggle over rights to biological resources, as the commercial potential of tropical biodiversity is realised. Now they see the threat of the ultimate exploitation, that of their own genetic material. Indigenous peoples' organisations are keenly aware that modern genetic research is in large part a quest for patent rights, and that the greatest of many controversies surrounding the HGD Project have concerned the issue of patenting human DNA sequences.

The data gathered by the HGD Project will form a species of race gallery, but the indigenous peoples' concerns are better expressed by the metaphor of a race library. This establishment will not only own the books, rare editions all, but the copyrights to them as well. And it may not be long before many of the samples become museum objects like the ancient skulls in the race galleries; the last traces of extinct peoples, preserved and organised so as to maintain their scientific utility.

Race galleries may also be established outside science. The most ambitious of all these is the one which, every ten years, attempts to sort all the 240 million citizens of the United States into racial categories. Although only two of the eleven censuses from 1890 to 1990 have used exactly the same set of categories, certain themes endure. The principal one is the distinction between White and Black. In 1890, census enumerators were charged with the task of classifying people of mixed African and European descent as Mulattos (half and half), Quadroons (a quarter black), or Octoroons (one eighth black). This proved completely unworkable, and the mixed categories were discarded in 1900; however, the Mulatto class reappeared in 1910 and 1920. From 1930, the 'one-drop rule' was subsequently adopted: 'one drop of black blood' consigned an individual to the Black category. Mexicans appeared as a separate category in 1930, but were subsumed into the White class in 1940.[15]

The Chinese, Japanese and American Indians have been listed separately throughout. Indians from India featured as 'Hindus', regardless of their actual religious affiliation, in 1930 and 1940, then disappeared until 1980, when they returned as Asian Indians. During the same period, the number of what can be termed Pacific categories has gradually risen, albeit with fluctuations from census to census. In the 1980s, bureaucrats concerned with ethnic classification instituted a supercategory called 'Asian or Pacific Islander'. This featured in the 1990 census as the umbrella for the categories of Chinese, Filipino, Hawaiian, Korean, Vietnamese, Japanese, Asian, Indian, Samoan, Guamanian and Other API. Outside the API class, the two indigenous peoples of Alaska – Eskimo and Aleut – were listed separately.

The US Department of Education came up with a somewhat more rococo version of this scheme to present to its job applicants. It included the category of 'Gypsy' among its nineteen ethnic codes, but despite defining this as 'a person having origins in the original Gypsy groups of Europe', placed it in the Asian or Pacific Islander section. 'East Indian' was defined as 'a person having origins in the original peoples of India, Ceylon and, in some cases, Pakistan'. Which cases were not specified, nor the category or categories a person of Pakistani origins should be assigned to in the other cases. The form also included an additional four classes under the heading of 'Latino', a super-category defined as encompassing various Spanish-speaking cultures 'regardless of race'. (It did, on the other hand, group the two native Alaskan populations together.)

In scientific as well as intuitive terms, this is a dog's breakfast. Criticising the 1980 census classification, the sociologist William Petersen trenchantly observed that it violated 'elementary rules for constructing a taxonomy – that the classes be mutually exclusive, that all the classes add up to the whole of the population, and that they be of roughly the same order of importance and magnitude'. The logical absurdity of the classification project is exposed by demands for a 'multiracial' category, open to all of 'mixed race' (the closest option in previous censuses has been 'Other'). These have been met with objections from African-Americans who point out that almost all their community is, in the sense of ancestry, multiracial.[16]

All race galleries, then, are vulnerable to accusations of structural weakness or incoherence, whether they are constructed on a biological or a sociological basis. The US bureaucratic galleries are essentially expressions of social and political forces; particularly the great narrative of white domination over black, and the tendency for interest groups to be formed on ethnic lines. But they have not totally expunged biology. Indeed, their obscurity and lack of definition help sustain it. On the census form, the term 'race' covers all categories, from the all-embracing 'White' to the hair-splitting Aleut. Such a usage was commonplace in the past, but now sits awkwardly with usages that have developed in response to the growth of unease about concepts of race and human group difference.

In some cultures, such as that of the equivalent bureaucracy in Britain, the word 'race' itself has been erased, and replaced by the term 'ethnic group'. This move was proposed in 1935 by Julian Huxley and Alfred C. Haddon in a book entitled *We Europeans*, aimed at a popular audience and intended to counter Nazi race science. They advanced the argument that has since become familiar as the basis of the scientific case against racism, that race is not a biologically meaningful concept in humans. In other species, the term 'race' had been replaced by 'subspecies'. The forms classified as subspecies were separated by relatively clear-cut lines between them, however, instead of shading into each other like human varieties. 'In the ultimate analysis . . .' Huxley and Haddon observed, 'the decision as to what is a "race" is a personal matter resting largely on subjective impressions'.

They did not, however, rule out the possibility that human races had existed at some stage in the past, before interbreeding muddled the distinctions between them. These hypothetical major races, corresponding to today's whites, yellows and blacks, would be better termed 'major subspecies' of the single species *Homo sapiens*. It was 'difficult to resist the conclusion' that there had once been geographical groups corresponding to the main European types, Nordic, Mediterranean and Eurasiatic (which included the Alpine): these they dubbed 'minor subspecies'.[17] Race was thus meaningless

23

today, but had probably been a real factor in the formation of living ethnic groups. *We Europeans* was a transitional work: in relegating race to history, it showed that it was still rooted in the race concept. The focus of its attack was not the idea of the Nordic type, but the Nordic myth as developed by the new National Socialist regime.

The theoretical difficulties of the race concept had been accumulating for many years, and were being compounded by the emergence of genetics as a new way of describing the human biological diversity. The suggestion of Huxley and his colleagues, that 'race' should be replaced by 'ethnic group', was most thoroughly implemented where 'race' was more or less synonymous with 'nation'. Speaking of a Welsh or French 'race', once a commonplace usage, is now archaic. But ethnic groups are understood to be culturally defined. Common usage rarely refers to 'whites' or 'Caucasians' as an ethnic group. These are terms for a category that is seen to belong to a different level in the classification system for human groups, and therefore requires a different name. So the term 'race' endures, implying that in some sense it is biologically meaningful.

The issue has remained ambiguous ever since. According to the US Department of Education system, it is possible to be a Latino 'regardless of race'. So race exists as a category distinct from the ethnic group. But its relationship to biology is unclear. The sociobiologist Pierre van den Berghe has suggested the distinction that races are socially defined upon physical criteria, whereas ethnic groups are socially defined on cultural criteria.[18]

This definition has won widespread acceptance in the social sciences. Yet van den Berghe is also an enthusiastic advocate of the argument that humankind is biologically predisposed to divide itself up into such groups, which is far less palatable to his sociological colleagues. This line of thought, deftly permitting race to be both subjective and rooted in human biological nature, is a significant development in the race theory of recent years.

Among the many poetic truths that have emerged from psychoanalytic theory is the notion that what has been repressed always returns.

Race is the great repressed of twentieth-century science. Conventional wisdom holds that, like Marxism in politics, race in science is an idea whose historical moment has passed. But, approaching the close of the century, it is important to examine the counterpossibility: that scientific anti-racism is a doctrine belonging to a historical phase which is now in its terminal stages.

The anti-racist model, the Unesco scheme of human diversity, was part of the new world order established after the defeat of Germany and Japan in 1945. That order had many other components, the greatest of them being the Cold War divide between the Soviet bloc and the Atlantic alliance. Parts of the post-war settlement were locally specific. In Britain, it took the form of a social-democratic consensus which accepted that the state would guarantee welfare and would own a significant proportion of industry. Up until the 1970s, all of these seemed like givens of existence. The British post-war settlement has been revoked; the overarching given of the Cold War has evaporated. If the Soviet Union itself could fall to bits, why should we expect the anti-racist element of the post-war order to last for ever?

There are two obvious answers, once scientific and one political. The scientific answer is that the return of racial science is objectively impossible, the evidence against race being too strong. The political answer is that history has now entered the post-colonial phase, with power no longer the exclusive preserve of Europeans, rendering impossible the return of racial science to the position of ascendancy in the world of ideas it once enjoyed.

These claims contain a great deal of force. But they may be missing the point. To illustrate their weakness, a useful analogy may be drawn with fascism. In order to demonstrate that fascism poses a threat in Western Europe today, it is not necessary to devise a plausible scenario for a Fourth Reich. It seems highly unlikely that a democratic state in the European Union should convert itself into a totalitarian one, whether by electoral or other means. The only domain in which fascism is capable of wresting hegemony from liberal democracy is that of fantasy, and even that is confined to thriller novels and the imaginations of the politically marginal. Yet fascism is undoubtedly a growing force in Western European politics today. It is able to influence the mainstream from the margins in a number of ways: by violence, by campaigning on issues such as immigration, and by commanding small but significant percentages of electoral support. Neo-fascists may never be able to take power in a European Union country on their own, but the example of Italy in 1994 showed that they may be able to govern as part of a coalition. Such elements have by definition entered the mainstream. Less tangibly, neo-fascist groupings have drawn the mainstream towards them, reshaping political issues in their preferred images.

Similarly, it seems most unlikely that the great nineteenth-century race gallery can ever be rebuilt. The historical moment of race science as a dominant system of belief has passed. But intense passions are invested in smaller, newer race galleries. A black American youth embraces the belief that melanin does not simply pigment his skin, but bestows psychic powers upon him and his fellow people of African descent. A white scientist devotes himself to the hypothesis that American children of East Asian descent do well in school because their distant ancestors evolved to cope with a particularly severe Ice Age. And a European museum curator struggles, largely successfully, to defend his hall of skulls and what they represent.

The Nazi regime, a totalitarian system of social order founded upon a system of biological order, implemented totalitarian racial policies which culminated in genocide. The Third Reich now stands as the first point of reference in any discussion of racial science. It could not and should never be otherwise. But the danger is that the scale of the monstrosity overwhelms the discussion, causing other possible ways in which race science could exert its effects to be overlooked. The great political ideologies of the earlier part of the century were peculiarly dangerous because they believed that they could utterly transform human society. The prevailing mood nowadays, by contrast, is fatalistic. Among the majorities, faith has ebbed – in ideologies, in the power of governments to ameliorate social ills, and in the power of people to change their lives by political action.

This loss of faith has been accompanied by its corollary, a shift towards the belief that human society is preordained to be the way it is; that nature shapes our lives more than nurture. Genetics is seen to be the dominant system of knowledge in the biological sphere, and increasingly provides society at large with models and metaphors for understanding the world. At the same time, the replacement of ideological divisions by ethnic ones has confirmed the sense that humankind is as permeated by consciousness of race as ever before. The idea that this is part of human nature is already influential. It may only be a matter of time before significant connections start to form between gene consciousness and race consciousness.

In that event, the greatest danger is not that new machineries of genocide will be established. The 'ethnic smart weapon', designed to kill only those of a particular race, will remain in the realm of science fiction. But fatalism may be just as lethal. If the urban crisis in America continues to worsen, white American suspicions that blacks are inherently criminal or uneducable will grow. Such ideas will increasingly obtrude into mainstream political discourse.

As they do, racial science will return as an ideology to legitimate these prejudices, and to justify the proposition that money spent on African-Americans is money wasted. Arthur Jensen first advanced such a claim, in respect of education, in 1969. This proposition has since developed into a theoretical industry, but its influence outside psychology has been limited by the political unacceptability of its message. Its sudden reappearance upon the publication of Richard Herrnstein and Charles Murray's *The Bell Curve* suggests that the race concept may now be poised at the threshold of acceptability.[19]

Anti-racist science is not nearly as ready to meet this intellectual challenges as it likes to make out. Nor will it be, until it treats the modern race galleries – whether of Vienna, or IQ tests, or Afrocentrism, or the Croatian frontlines – with the seriousness they deserve.

2

Applied Biology

THE CLAIM THAT there is nothing new in race science, a perennial cry of the second half of the twentieth century, and quite probably much of the twenty-first, finds support from the founding fathers of the tradition. The racial term 'Caucasian', now used to denote the peoples who extend westwards from India to Europe and the northern shores of Africa, was coined in the late eighteenth century by Johann Frederich Blumenbach of Heidelberg. It expressed a belief in ideal type which was closely allied to aesthetic perceptions, apparent in Blumenbach's reference to 'the most beautiful race of men, I mean the Georgian'.[1] The ideal was intimately linked to the original: 'if it were possible to assign a birth-place to the human race all physiological reasons would combine to indicate that place . . .'

Blumenbach proposed four other races; the Mongolian, the Malay, the Ethiopian and the American. These he considered to be modifications of the Caucasian. 'The skin of the Georgians is white and this colour seems to have belonged originally to the human race,' he observed, 'but it can easily degenerate to a blackish hue.' This view was consistent with the longstanding European cultural tradition that gave positive connotations to lightness and negative ones to darkness. More fundamentally, it refracted the Christian tradition of a single origin of humanity through the optimistic and inclusive vision of the Enlightenment, upholding the ideal of an essential human unity.

This was underscored by Blumenbach's acknowledgment of the basic problem of human taxonomy, the breadth of human diversity and its continuous variation:

Although there seems to be so great a difference between widely separate nations, that you might easily take the inhabitants of the Cape of Good Hope, the Greenlanders and the Circassians for so many different species of man, yet when the matter is thoroughly considered, you see that all so do run into one another, and that one variety of mankind does so sensibly pass into another, that you cannot mark out the limits between them.

What could very definitely be marked out, however, was the gap between humans and the rest of Creation. Carolus Linnaeus, founder of the Latin binomial system of taxonomy, was one of many who were apt to suspect that different types of human might have had different origins, and that some were closer to animals than others. Reports reached Linnaeus that the French experimentalist Réaumur had succeeded in crossing a rabbit and a chicken, producing chicks with fur instead of feathers. 'With regard to the human race,' Linnaeus remarked, 'one might be induced to think that Negroes have a rather peculiar origin but for my part I refuse to believe this.'[2]

During the first half of the nineteenth century, the intellectual climate encouraged those inclined to believe in separate origins, or polygeny, to become bolder. This was the period in which, through its expanding consciousness of time, science became irreconcilable with a literal interpretation of Scripture. It was realised that the Earth was far older than the few thousands of years proposed by Biblical exegesis, and that fossils were the remains of species which had come and gone within the vast span of geological time. The idea of the Great Chain of Being, a hierarchy of living forms with Man at the top of the material order, could now be conceived chronologically, with increasingly sophisticated forms superseding each other through the eons. 'Life', as Nancy Stepan puts it, 'had a progressive history.'[3]

The white race was allotted the foremost place in life's advance, and as science detached itself from the hegemony of the literal Bible, it became easier to propose closer relationships between other races and other species. Polygeny also guaranteed the idea of racial purity. If different stocks were taken to have had completely separate origins, then they must have originally existed in a pure and distinct condition. Monogeny was somewhat harder to reconcile with racial purity, but geological time resolved the difficulty: the separation of the races was taken to be so ancient that it was as if they had never been one.

In parallel with the concept of race, the idea of *Volk* gained ground. Its principal author was Joachim Gottfried von Herder, who shared Blumenbach's humanist disposition, but embodied the Romantic belief that there were fundamental truths that could not be encompassed by Enlightenment rationalism. Romanticism did not wish to subject the spirit unique to each people, or *Volksgeist*, to scientific inquiry. Yet its insistence on the importance of what we would now call ethnic identity resonated with the growing importance attributed by science to the race concept. History came to be understood as the history of racial struggle; anthropology, the science of man, was primarily the science of racial variation. Even class relationships were readily understood as racial ones. The French Revolution was conceived as the revenge of the Gauls upon the ruling Franks; the Comte de Gobineau considered that the common people 'belong to a lower race which came about in the South through miscegenation with the Negroes and in the north with Finns.'[4]

The focus of anthropological inquiry was the skull, seat of the intellect and therefore of man's defining characteristic. It endures after death, usually longer than any other skeletal remains, and it was taken to be marked by enduring indicators of racial identity. To derive the meaning of these signs, the skull had to be placed in the matrices of systematic knowledge on which science was learning to base itself. The number of parameters grew to be legion – and many of them flourish to this day – but the two definitive measures are the facial angle and the cephalic index.

The facial index was devised in the 1770s by Petrus Camper, a Dutch anatomist and painter, who sought to establish a geometry of beauty.[5] One line was drawn across the profile from the base of the nose to the earhole, and a second from the upper lip (or the edge of the front teeth in skulls) to the forehead. The angle between the two passed beyond the perpendicular in Greek statues, which Camper found to be fashioned with reference to an ideal of 100°. It diminished along a series which corresponded to the Great Chain of

Being, from the European to the Negro to the monkey to the dog and eventually, as Camper noted, to the woodcock. Facial projection, or prognathism, came to be understood as a cardinal sign of primitiveness, as in Thomas Huxley's much-quoted observation that 'it is simply incredible that . . . our prognathous relative . . . will be able to compete successfully with his bigger-brained and smaller-jawed rival, in a contest that is to be carried out by thoughts and not by bites'. He was speaking of the Negro, but prognathism was also identified in the Irish, encouraging cartoonists to depict Irishmen as resembling gorillas, and in women. Facial projection was thus a reliable marker on the Great Chain of Being.[6]

The cephalic index is the breadth of the skull of a living person, divided by the length and multiplied by a hundred; when applied to dead skulls, it is called the cranial index. Introduced by the Swedish anatomist Anders Retzius in the 1840s, it provided the measure which defined racial studies for the rest of the century, as they became preoccupied with the racial history of Europe. The division of European crania into dolichocephalic, or long-headed, and brachycephalic, or broad-headed, was taken to reveal the existence of primordial races whose balance was shifted by invasion and conquest.

By the end of the century, the tripartite scheme of Nordic, Alpine and Mediterranean had emerged. These were not simply physical types. They were imbued with souls, each with a peculiar *Volksgeist* contained within its characteristic crania. The symbiotic relationship between taxonomy and Romanticism proved immensely influential, remaining commonplace wisdom until the period between the world wars. One commentator of that time attributed a 'plodding, detail-loving industry' to the short-headed Alpines.[7] Another observed that 'it is in the lighter branches of engineering, electrical and motor-car work that the Mediterranean is at home, and the modern garage staff is composed almost entirely of men of this race'.[8]

Physical type and racial character were forcibly separated by the Second World War and its legacy, but the belief in the reality of racial type proved far more durable, at least in certain quarters. Its persistence is a powerful testimony to that mode of Western thought which creates an arbitrary opposition between two categories, and invests them with the profoundest significance. Skulls vary continuously in shape; Nature does not draw a line between the long and the broad. Even with modifications, such as the intermediate 'mesocephalic' class, the system is that of a binary opposition which is assumed to be a dependable indicator of relationship. Within Europe, the assumption was at least plausible. People had to come from somewhere: why should the similarities between living Welsh and Southern Europeans not be taken as evidence that they derived from the same Mediterranean race? The fact that long heads and broad heads could be found in the same family was taken not as a sign of the range of variation between individuals, but of the fundamental strength of racial type, which reappeared down the generations to affirm the various original stocks that had combined in the family's ancestry. Race, like blood, would out.

The task of the scientist, as George Stocking has pointed out, was to see through the variation to the underlying essence. Within the discipline, however, there was an underlying unease about the dependability of the skull. While in one instance it supported no fewer than 5,000 measurements, the case which epitomised the problem was that in which the German anthropologist Otto Ammon, trying to respond to a request from William Z. Ripley, failed to find a single photograph of a 'pure' Alpine type. This, Stocking notes, was despite the subjection of 25 million Europeans to anthropometric measurement over the last third of the nineteenth century. In 1911 came the pivotal – though not universally accepted – demonstration that skulls were plastic, when Franz Boas showed that the heads of the children of immigrants to the United States differed in shape from their parents'.

In a sense, race was everything because science had nothing more convincing at its disposal. The idea of evolution arose before Darwin and became a constant, but an uncertain one. Darwin's great insight – grasped at the same time by Alfred Russel Wallace – was that some organisms within a population leave more offspring than others because their characteristics make them better fit for survival. If these characteristics are inherited, they will thus tend to accumulate in successive generations, by what Darwin termed 'natural selection'.

This seemed too negative a force to many scientists, however; its stock fell especially low around the turn of the century.

Nor did the rediscovery of Gregor Mendel's experiments on the transmission of characters down the generations immediately transform the situation. In the 1860s, crossing strains of peas, Mendel had worked out that the colour of peas was determined by pairs of factors, which might be dominant or recessive, but do not blend together. Mendelian inheritance appeared to be a conservative agency, preserving the old rather than generating novelty. Although the idea of change through genetic mutation was proposed in 1900, the elements necessary for a comprehensive theory of population genetics (including the insight that mutations could be produced by environmental factors such as radiation) were not assembled until the 1930s. By that time, other matters were pressing.

The strong German state originated with Bismarck, who also pioneered the political manipulation of hostility to Jews by encouraging anti-semitic groups. Darwin's cousin Francis Galton created the doctrine of eugenics, the Social Darwinist doctrine that active measures should be taken to improve the quality of the racial stock by influencing the breeding rates of different sections of the population. Joachim Herder established the principle of *Volksgeist*, the immanent soul of a people. A series of authors, most notably the Comte de Gobineau and Houston Stewart Chamberlain, had taken the observation that European languages resemble Sanskrit, and turned it into the myth of an Aryan people who had imposed civilisation upon Europe. In 1900, the military industrialist Friedrich Alfred Krupp funded a prize for an essay competition on the question 'What can the theory of evolution tell us about domestic political development and the legislation of the state?' One of the judges was the *völkisch* Darwinist Ernst Haeckel, a proselytiser for biological determinism whose scientistic tract *Welträtsel*, (Enigma of the World) sold half a million copies. At the level of ideas, the National Socialists had very little new to do, except make the connections between already thriving traditions, and mobilise them within a totalitarian project.

Rhetorically, one of Adolf Hitler's most powerful devices was the relentless insistence on the imperative; expressed most succinctly and forcefully in the notion of the Will. Charles Darwin had predicted that 'the civilised races of man will almost certainly exterminate, and replace, the savage races throughout the world'. Hitler declared that 'the *völkisch* concept separates mankind into races of superior and inferior quality. On the basis of this recognition it feels bound, in conformity with the eternal Will that dominates the universe, to postulate the victory of the better and stronger and the subordination of the inferior and weaker.'[9] Nature was imbued with an uncontradictable purpose, expressed in biological laws. Thus it was an 'iron law of nature' that animals mate only with their own kind; hybrids were frequently sterile and always inferior to pure stock: 'Such mating contradicts the will of nature towards the selective improvement of life in general.'

Hitler believed that there were inherent limits to the power of the state. The idea that man could control Nature was a 'piece of Jewish babble'. Rather, the state existed to fulfil racial destiny, and this was above all a biological duty; any states which did not exist for this purpose were 'monstrosities'. The lessons Hitler drew from history was that 'all the great civilizations of the past became decadent because the originally creative race died out, as a result of contamination of the blood.' The most creative of all races was the Aryan, the 'archetype' of Man, responsible for 'every manifestation of human culture, every product of art, science and technical skill'. It was necessary for Aryans to find a suitable environment, in the form of a congenial climate such as that of Greece, for their creative faculty to find expression: the cold of the north had inhibited the early Germans. But as the creative faculty was exclusive to the Aryans, the Laplanders or Eskimos would not have created a culture if they had wandered south to Greece.

The overriding necessity for the German state, therefore, was to nurture the Aryan elements within the German people. Few political jokes can have circulated so widely as the jibe that the ideal Aryan was as blond as Hitler, as long-headed as Göring, and as tall as Goebbels. The Nazis were hardly unaware of the problem themselves, though. 'Unfortunately the German national being is not based on a uniform racial type,' admitted Hitler. He blamed immigration and the openness of German borders, portraying the

continued existence of different types as a biological threat to
political unity:

> Beside the Nordic type we find the East-European type, beside
> the Eastern there is the Dinaric, the Western type intermingling
> with both, and hybrids among them all. That is a grave drawback
> for us. Through it the Germans lack that strong herd instinct
> which arises from unity of blood and saves nations from ruin in
> dangerous and critical times.

At such times, by contrast, 'the Germans disperse in all directions'.

The Nazis' totalitarian efforts to develop a herd instinct, epitomised by the Nuremberg rallies, can thus be seen as an environmental
intervention to counter what they understood as a biological
weakness. But even a racial cloud may have a silver lining. Hitler
suggested that the failure to form a new race had permitted the
survival of pure Nordics, Germany's 'most valuable treasure'. This
perception informed his definition of 'the supreme purpose of the
ethnical State': 'to guard and preserve those racial elements which,
through their work in the cultural field, create that beauty and
dignity which are characteristic of a higher mankind.' The state was
conceived as fundamentally biological and organic; not 'a piece of
mechanism alien to our people, constituted for economic purposes
and interests, but an organism created from the soul of the people
themselves'.

This was the ultimate extremity of Romanticism, a vision of a state
that was a 'living organism', despite its commitment to technology
and modernisation, because of its racial nature. Like many of his
followers, Hitler did not consider that speaking of the soul
invalidated scientific pretensions. His ideology was 'no mystical
doctrine, but rather a realistic doctrine of a strictly scientific nature'.[10]

Naturally enough, scientists themselves enjoyed opportunities for
influence during the National Socialist period, provided that they
worked in appropriate disciplines and were not Jewish. The racial
duty of the state had to be implemented through the integrated
operation of legal measures, medicine, scientific research, and
bureaucratic procedures by which genealogical and reproductive

aspects of the population could be monitored. Many scientists had been supporters of the international eugenic movement before the National Socialists came to power, and they welcomed the advent of a regime committed to eugenic principles. *Rassenkunde* (racial anthropology) and *Rassenhygiene* (racial hygiene) became privileged fields of knowledge.

One of the most prominent beneficiaries of the new order was Hans F. K. Günther, known as '*Rassen*-Günther', a self-taught physical anthropologist who wrote a number of popular books in the 1920s on Nordic romantic themes, as well as the influential *Rassenkunde des deutschen Volkes*. When the Nazis entered the regional government of Thuringia in 1930, the minister of education appointed him to the University of Jena, despite the opposition of most of the university's senate. His inaugural lecture was attended by Hitler himself; it was around this time that Günther joined the Party. As well as popularising *Rassenkunde* and the idea of Nordic supremacy, he contributed to the problem of how to identify Jews by describing 'typical Jewish posture'.[11] Much later, in 1944, he undertook to give a lecture on 'The Invasion of the Jews into the Cultural Life of the Nations' to the International Anti-Jewish Congress, but the meeting was cancelled because of Germany's deteriorating military fortunes.[12]

Günther took practical steps to promote racial purity too, collaborating with the racial hygienist Fritz Lenz and the Gestapo to organise the secret sterilization of the *Rheinlandbastarde*, children fathered by black French occupation troops in the Rhineland after the Great War. And he put figures on Hitler's racial anxieties, calculating that a mere 6–8 per cent of the German people were pure Nordics. Pure Mediterraneans made up 2–3 per cent; similar proportions were pure Alpines and Baltics. In the face of his own statistics, he called for a 'Blond International'.[13]

Otmar, Baron von Verschuer also flourished under the Nazi regime. His specialism was the relationship between heredity and environment, which he investigated by means of studies on twins, the 'sovereign method for genetic research in humans'. He also proselytised for the united cause of genetics and National Socialism, although he did not join the party until 1940. According to von

Verschuer, genetics revealed the biological links between individual, *Volk* and race. In medicine, the individual was no longer to be treated as an individual, but as 'one part of a much larger whole or unity: his family, his race, his *Volk*'. As well as campaigning for increased attention to racial hygiene on medical school and university syllabuses, von Verschuer established a journal, *Der Erbarzt* (The Genetic Doctor), 'to forge a link between the ministries of public health, the genetic health courts, and the German medical community'.

Speaking in 1939, he claimed that 'we geneticists and racial hygienists have been fortunate to have seen our quiet work in the scholar's study and the scientific laboratory find application in the life of the people'. Von Verschuer was not merely a scientist who made the most of the fact that his line of research had come into political favour, he was a racial activist committed to the efficient functioning of organs within the body of Hitler's 'ethnical State'.

Like any scientist, however, von Verschuer was keen to keep up his research, and during the war he complained that the supply of 'twin materials' had been disrupted by the hostilities. But the war also provided new sources of 'materials' in the concentration camps, and the state provided funds to process the data gathered. When a former graduate student, now in the SS, asked his mentor for career advice, von Verschuer told Dr Josef Mengele to take up a post at Auschwitz, a site which offered 'unique possibilities' for research because of the 'diverse racial groupings' assembled there.[14] Mengele acted as an assistant to von Verschuer at the Kaiser Wilhelm Institute for Anthropology, sending various biological materials from Auschwitz to the Institute. These included pairs of eyes from twins Mengele had dissected after their murder, blood samples from twins he had infected with typhus, children's internal organs, corpses, and the skeletons of murdered Jews.

After the war ended, von Verschuer denied receiving these specimens, and claimed that he was 'openly opposed to the National Socialist race fanaticism'. To restore the reputation of science, he recommended the purging of those who were 'not real scientists'. The denazification authorities classified him as a 'collaborator', and fined him 600 marks. He was unsuccessful in his attempts to get the

Kaiser Wilhelm Institute for Anthropology re-established, however. According to the head of the commission overseeing the reconstruction of the Kaiser Wilhelm *Gesellschaft*, 'Verschuer should be considered not a collaborator, but one of the most dangerous Nazi activists of the Third Reich. An objective judgment of the investigative committee must recognise this, and thereby take actions to guarantee that this man does not come into contact with German youth as a university teacher, or with the broader population as a scientist in the fields of genetics and anthropology.' Despite this verdict, von Verschuer was appointed professor of human genetics at the University of Münster in 1951.

Hans Günther fared somewhat less well than von Verschuer and other of his colleagues in racial hygiene, failing to gain a new academic post despite having been cleared by the denazification courts. One suggested reason was the abstruseness of his thought; another was a personality described as 'almost antisocial'.[15] Nevertheless he continued to publish books – sometimes under pseudonyms – and he retained a dedicated following in certain circles. In his memoir *Mein Eindruck auf Hitler*, he claimed that he had found party politics 'tasteless and dirty' since 1919.

Although the protestations of scientific disinterest were clearly egregious in these two cases, they were not untypical of the line taken by German scientists in general. The historian Robert Proctor notes a greater readiness to complain about the disruptive effects of the war than to express remorse or grief, 'or any of the kind of soul-searching one might have expected after such a history'.[16] Bemoaning the intrusion of politics into science was a popular device.

Another technique for exculpation draws attention to the quantity of 'acceptable' science performed during the Nazi period. The physical anthropologist Ilse Schwidetzky, writing in the early 1990s for a series of historical papers edited by the British geneticist Derek F. Roberts, observes that anthropological journals such as *Zeitschrift für Rassenkunde* were 'filled predominantly with normal unpolitical studies'.[17] But the *NS-Zeit*, or National Socialist time, was not a normal period. Nazism was a totalitarian ideology which demanded that science serve political goals, and one of its most intimate intellectual relationships was with race science. Schwidetzky's own article on maps of physical traits in '*Mitteleuropa*', for example, published in *Zeitschrift für Rassenkunde* in 1937, could not help but be political.[18] Nor could her other contributions, such as several reviews of books on the 'Jewish question', particularly that of *World Struggle. The Jewish Question Past and Present*. It was published in Frankfurt in 1941 by the Institute for Research into the Jewish Question, a body headed by Alfred Rosenberg, whose name mocked his extreme anti-Semitism.[19] By the time Schwidetzky's notice appeared, the National Socialist regime's chosen solution to the 'question' was being implemented.

The assertion of scientific neutrality is not simply in the interest of German scientists whose career overlapped the National Socialist period. Nazi science is the definitive retort to the claim that science is 'value-neutral'. Since the involvement of scientists in the National Socialist project is an undeniable matter of historical record, to uphold the model of neutrality demands that this involvement be seen as an aberration. But Nazi science was solidly rooted in traditions that were not confined to Germany and were not radically divergent in quality, rather than intensity, from their counterparts in other countries. Although many scientists outside Germany had become dissatisfied with race as a means of classification, and grew increasingly troubled about the use the Nazi state made of science, they did not make a radical break with the race concept until after the war.

By contrast, it may be appropriate to describe the machinations of Stalin's favourite, Trofim Lysenko, as a distortion of science, since he used the Soviet apparatus to replace the internationally accepted theory of genetics with a pseudoscientific doctrine of his own which flattered the tenets of Marxism-Leninism. If what the Nazis did with race science constituted a distortion, this lay solely in the extremity to which they took it. Their own view, originally formulated by Fritz Lenz, was that 'National Socialism is nothing but applied biology'.[20]

After the Allied victory, German scientists rushed to reassert the natural separation of science and politics, or to pledge allegiance to the new order. In some cases, political revisions could be effected by the academic equivalent of taking down the portrait of the old ruler and replacing it with that of the new one. Robert Proctor illustrates a shameless example by comparing the 1938 and the 1947 editions of Paul Diepgen's *Die Heilkunde und der ärztliche Beruf.* Among several endorsements of National Socialist values in the earlier edition is an allusion to 'the National Socialist *Weltanschauung*'. In the post-war edition, this has been replaced with the sentiment that 'Science serves the whole world and must be cosmopolitan'.[21]

This piece of cut-and-paste revisionism caught the note exactly, as did *Zeitschrift für Rassenkunde*'s change of name to *Homo*. (It has continued into the 1990s with this title, under Ilse Schwidetzky's editorship.) After a world war, peacemaking and reconstruction had to be undertaken by the world as a whole. The humanist, universalising tradition of Enlightenment science had to be rediscovered, and applied at both the practical and philosophical levels to the rediscovery of human unity. The newly dominant opposition was not between Nazism and democracy, or between fascism and communism, but between the genocidal divisiveness of the Third Reich and the universalistic idealism of the United Nations era. An attractive aspect of the new vision, not least for Germans with a past to put behind them, was that it seemed to elevate science above politics.

The ethical code of this new world was promulgated in 1948. It was nearly called the International Declaration of Human Rights, but René Cassin proposed an amendment at the UN General Assembly substituting 'Universal' for 'International'. The latter refers to a particular set of historical circumstances, to the sets of relationships which reconcile the particular interests of the actors that happen to be on the world stage at a given moment. Universal Man is above all that. And in Donna Haraway's words, 'natural science would be needed to get post-World War II universal man off the ground'.[22]

Science fell within the remit of Unesco, the United Nations Educational, Scientific and Cultural Organisation, established in 1945. Its constitution said that the war had been made possible by the 'doctrine of the inequality of men and races'. In 1949, it began discussions on science and race, in response to a request from the United Nations Economic and Social Council that it consider 'a programme of disseminating scientific facts, designed to remove what is generally known as racial prejudice'.[23]

The Statement on Race that emerged from Unesco House in July 1950 launched straight into the universal. 'Scientists have reached general agreement that mankind is one: that all men belong to the same species, *Homo sapiens*,' it began. Within the species, group variation was defined not by type but by populations, 'each one of which differs from the others in the frequency of one or more genes'. The total number of genes was vast; the number of genes whose varying frequencies defined populations was small by comparison, and therefore 'the likenessses among men are far greater than their differences'. A race might be defined by a concentration of genes or physical traits, but its significance was subjective. 'What is perceived is largely preconceived, so that each group arbitrarily tends to misinterpret the variability which occurs as a fundamental difference which separates that group from all others.'[24]

The problem with the term 'race', however, was the looseness of popular usage, which still spoke of an English or a Chinese race. Repeating the recommendation made by Huxley and Haddon in *We Europeans*, the Statement on Race proposed replacing the term with 'ethnic group'. This was interpreted by some as denying the reality of race, but with hindsight the move appears conservative, saving the biological reality of race from the muddle of folk usage. All that could generally be agreed at the biological level, however, was the tripartite division into Mongoloid, Negroid and Caucasoid; no such consensus existed about further subdivision.

'Whatever classification the anthropologist makes of man,' it continued, 'he never includes mental characteristics as part of those classifications. It is now generally recognised that intelligence tests do not in themselves enable us to differentiate safely between what is due to capacity and what is the result of environmental influences, training and education. Wherever it has been possible to make allowances for differences in environmental opportunities, the tests have shown essential similarity in mental characters among all human groups.' The evidence indicated that cultural history rather than genes was the major factor behind differences in cultural achievement. 'The one trait which above all others has been at a premium in the evolution of men's mental characters has been educability, plasticity . . . It is, indeed, a species character of *Homo sapiens*.'

This was fundamental to the post-war anti-racist vision of human evolution. The defining step in the emergence of modern humans had occurred in the brain, creating a mental engine with global capability. It was a general processor akin to the universal computer whose theoretical properties were set out during this period by the mathematician Alan Turing. Instead of operating according to a principle of sufficiency, through specific adaptation to a given environment, it gained its strength from excess, which permitted humans to impose gross modifications upon their environments, rather than the other way around.

With mental overcapacity as the dominant factor in the human biological condition, the role of evolution as a source of variation within the species was greatly diminished. The brain's expansion had effectively placed it beyond the reach of environmental pressure. For example, fire, clothing and artificial shelters were the primary means by which humans adapted to cold climates; physical adaptations such as a stocky build, while significant, were secondary.

The research sites favoured by the population geneticists for the study of human variation were islands or other places in which small human groups had become isolated. There they were subjected to the quartet of forces identified as driving population dynamics in isolated groups: drift, founder effect, mutation and selection.

Genetic drift is the tendency for the genetic composition of a small population to be altered by random events, such as an accident which wipes out a family. Founder effect is the influence of a particular random event: the formation of a new population from a small number of individuals. The smaller the group, the smaller the range of genes it will contain, and the greater the chances that two copies of a particular gene will come together in future matings. Many inherited diseases are carried by 'recessive' genes, which are only expressed when two of them pair up; small populations are thus likely to show higher incidences of such conditions than large ones. Through the process of mutation, novelty arises in the gene pool; through selection, the frequencies of genes old and new are adjusted to surrounding conditions.

Visible racial differences were less easy to investigate, but readily lent themselves to evolutionary explanations. It has been suggested that the thin noses found among European and Arctic peoples serve to moisten air and warm it up on its way to the lungs. Pale skin could be seen as an adaptation facilitating vitamin D synthesis in cloudy climates, by allowing ultraviolet light to penetrate the skin, while dark skin could be interpreted as an adaptive barrier to excessive ultraviolet radiation. Natural selection is not concerned with health hazards in themselves, though, but with how far they affect the number of offspring an individual produces. Protection against melanoma may not be enough to explain the prevalence of black skin. The disease is not usually fatal until its victim has reached adulthood, so although it could reduce the number of children such individuals have, it would rarely prevent reproduction altogether. And the darkest-skinned people do not all live in the sunniest regions: many of them live in parts of Africa that are rainier than Ireland. There remains the possibility that skin colour confers some other sort of selective advantage. A list of suggested functions compiled by Jared Diamond includes protection against heat, cold, frostbite, rickets, folic acid deficiency and beryllium poisoning. Diamond considers sexual selection – of traits found sexually attractive – to be a more likely explanation for skin colour than natural selection; as did Darwin, who concluded that 'not one of the external differences between the races of man are of any direct or special service to him.'[25]

Arguments like these, concentrating on the visible differences between groups, create a powerful impression of the superficiality of race. It is 'literally' only skin-deep, according to both Stephen Jay Gould and the psychologist Steven Pinker.[26] This is plainly untrue at the most obvious level. Norwegians and Nigerians would be easily distinguishable even if they were both emerald green in colour. At a more significant level, the claim implies that adaptation is skin-deep; that the external coating of the human body is the only part that varies according to different selection pressures in different regions. There is in fact evidence (discussed later in this account) that peoples originating in different climatic regions may differ in certain aspects of their physiology. But even if this is admitted, the brain still seems

43

to stand aloof. Popular accounts of science may thus give the idea that body and mind are separate.

It can be argued that anthropology has never come to terms with Darwinian evolution. According to George Stocking: 'Once beyond the emergent evolutionary moments of human cultural nature, antiracialist evolutionary paleoanthropology may be interpreted as essentially a superimposition of cultural determinist assumption on the later 99 per cent of human history; after evolution produced the one, culture produced the many.'[27] Evolution established a biologically level playing field for humankind; inequalities – or just differences – are the product of culture, and therefore beyond the explanatory powers of biological science.

As Western society, especially in the United States, has become increasingly preoccupied with economic flexibility and individual achievement, its need to believe in plasticity and overcapacity increases. In the colonial era, whites and blacks – for black and white, of course, is what this is all about – lived in different societies. One of the principal dynamics of the post-colonial world is human mobility on a mass scale. If our brains are fully plastic, our ethnic origins make no genetic difference to our ability to participate in whatever culture surrounds us. All human groups should therefore be, biologically speaking, equally valuable and transferrable 'human resources' for the global economy.

Meanwhile, the cultural need to deny the existence of limits to personal fulfilment has produced the folk belief in untapped mental potential, which offers a multicultural society the reassurance that all races are underachievers. It is expressed in the saloon-bar commonplace that 'we only use a third of our brains', to which the reply is, after a lengthy pause, 'I wonder what happens to the other third?'

On the question of psychology, the Unesco Statement observed that any group differences in temperament were probably overridden by individual ones, that personality and character were raceless, and that there was no biological justification for banning marriage between different races. It concluded with the assertion that 'biological studies lend support to the ethic of human brotherhood; for man is born with drives toward co-operation.'

The remarks about character and interbreeding speak more to the concerns of the first half of the century, but the propositions of the Statement as a whole are recognisable as the basis of the present-day liberal consensus on race. They are not, however, an authoritative intervention on the part of biological science. The committee was composed mostly of sociologists and social anthropologists, including Claude Lévi-Strauss; the Statement itself was largely the work of the rapporteur Ashley Montagu, who revised it in the light of criticisms submitted by a wider group of scholars, including the geneticists Julian Huxley, Herman Muller and Theodosius Dobzhansky. Its emphasis on culture represented the vigour of cultural anthropology, which flourished as an autonomous domain thanks in large part to Franz Boas's drive to separate biology and culture. Robert Proctor suggests the Statement 'represented the triumph of Boasian anthropology on a world-historical scale'.[28] One indicator of this success, noted by George Stocking, was that by mid-century, 'it was a commonplace for educated Americans to refer to human differences in cultural terms, and to say that "modern science has shown that all human races are equal".'[29]

What modern scientists actually believed, by and large, was that there was insufficient evidence to justify claims of racial superiority and inferiority. Of 106 geneticists and physical anthropologists asked to comment on the Statement, 80 responded; of these 23 accepted it as a whole, 26 agreed with the general spirit but took issue with particulars, and the rest strongly disagreed with it.[30] This last group included five out of six German respondents; among them was Fritz Lenz, who had been reappointed professor of human genetics at the University of Göttingen. He claimed that genetics had begun to destroy 'the fallacious concept of the equality or similarity of all men and the current belief in the omnipotence of social influences'. For good measure, he added that 'if an unprejudiced scientist were confronted with a West African Negro, an Eskimo and a North-West European, he could hardly consider them to belong to the same species.' Hans Weinert, of the University of Kiel, wondered 'which of the gentleman who signed the Statement would be prepared to marry his daughter to an Australian aboriginal'.[31]

A number of senior figures outside Germany were not yet prepared to endorse a position of agnosticism about mental differences. Herman Muller commented that he agreed with the intent to show the relative unimportance of genetic differences in comparison with environmental differences. 'But in view of the conspicuous hereditary differences,' he added, 'it would be strange for there not to be similar mental ones between averages.'

Muller also reckoned that the geneticists who had been consulted were more extreme in their environmentalism than most. The historian William Provine thinks he was right: the majority believed that mental differences existed, but had not yet been proved. Some went further, like Ronald A. Fisher, whose suggested revision stated that 'Available scientific knowledge provides a firm basis for believing that the groups of mankind differ in their innate capacity for intellectual and emotional development, seeing that such groups do differ undoubtedly in a very large number of their genes.'[32]

Dissatisfaction with the Statement, which was publicly aired in a lengthy correspondence in the British journal *Man*, published by the Royal Anthropological Institute, led to the convening of a new panel under a new rapporteur, Leslie C. Dunn. Its members, which included Dobzhansky and J.B.S. Haldane, were selected to give the revised Unesco position the biological authority that the first version had conspicuously lacked. The geneticists and physical anthropologists who now dominated the discussion upheld the basic elements of the argument, while discarding Montagu's idealistic declaration that 'mankind is one', opening with the plainer statement that 'scientists are generally agreed that all men living today belong to a single species . . .'

On the question of mental differences, the biologists paid little attention to the niceties of cultural relativism, observing that non-literate people perform worse in intelligence tests than 'more civilized' people. While moving to a position which acknowledged that innate differences might exist, the 1951 Statement compensated for this by making more than its predecessor of the possibility that biological variation within groups may be greater than that between them. Never mind the mean, it suggested, look at the range: 'within different populations . . . one will find approximately the same range of temperament and intelligence'.

The affirmation of the race concept was more explicit than before.

'We were . . . careful to avoid saying that, because races were variable and many of them graded into each other, therefore races did not exist,' Dunn noted in his commentary. 'The physical anthropologist and the man in the street both know that races exist; the former, from the scientifically recognizable and measurable congeries of traits which he uses in classifying the varieties of man; the latter from the immediate evidence of his senses when he sees an African, a European, an Asiatic and an American Indian together.'

This equation of the expert gaze of the scientist and the casual glance of the man in the street, a remarkably relaxed position for a scholar to take, is all the more striking for coming from a geneticist. Since then, traditional physical anthropology has become a scientific backwater. External appearances are regarded as an illusion; true wisdom lies in seeing beyond them, and beyond the superficial markers of race, to the genetic essence. But at mid-century, geneticists were rather more modest.

This was the historical moment at which genetics and physical anthropology formed a coalition around evolutionary principles. In 1950, the year that the first Unesco Statement based scientific anti-racism on evolutionary biology, three highly influential books appeared: William C. Boyd's forthrightly anti-racist *Genetics and the Races of Man*; *Races: a Study of the Problems of Race Formation in Man* by Carleton Coon, Stanley Garn and Joseph Birdsell; and *Origin and Evolution of Man*, the proceedings of the Cold Spring Harbor Symposium on Quantitative Biology.[33] This volume announced a populational and evolutionary 'New Physical Anthropology'.

If science really did proceed on the value-neutral lines of its ideal, these developments would have been a pure expression of the progress of scientific theory in the late 1940s. 'Palace histories of human genetics have it that the collapse of racialist thinking among biologists and anthropologists was due to the triumph of "popula-tional thinking" over "typological thinking" in the 1940s and 1950s,' observes Robert Proctor, noting a claim that this shift caused the collapse of the consensus that miscegenation was unhealthy. 'It is true that in the 1930s a number of geneticists began to realise that heredity was more complex than had been previously thought. As Provine has shown, however, there was no new scientific evidence introduced in the crucial period between 1939 and 1949, the period of decisive shift in attitudes towards race in most western countries.'[34]

The equation between populational thinking and anti-racism was also questioned by the typological rearguard, who denied the converse proposition that typology was automatically racist. A debate on the subject arose by accident in the early 1960s, when the journal *Current Anthropology* received two papers from Poland, one typological and one populational. Its editor, Sol Tax, decided to publish them together and give them the '*CA* star treatment', in which commentaries and the authors' responses followed the papers; the assembled articles appeared in 1962.[35] A rather plaintive accompanying note from Tax portrayed an anthropologist caught on the horns of cultural relativism. He had removed personal remarks from one Polish scholar's commentary, only to be told that the excisions were unacceptable. Tax climbed down, accepting that the Continental style of debate took sarcasm for granted. After the editor had made his final decision, however, another Pole warned Tax not to assume Dr Michalski's style was typical of the Polish academic community. None the less, Dr Michalski was permitted to compare Dr Bielicki to 'a homeopath amateur writing a specialized treatise in the field of surgery'.

At stake was Polish pride in a national school of typology based on mathematics, a field of scholarship in which Poland has traditionally excelled. This school of thought had been insulated from international developments by the Cold War, which impeded the circulation of people and papers between the Western and Soviet blocs. It was only in the late 1950s that younger anthropologists, among them Tadeusz Bielicki, began to absorb the ideas of the 'New Physical Anthropology'. Bielicki wrote the populationist paper; the typological tradition was upheld by Andrzej Wierciński. Bielicki later summarised the debate: 'On the whole, the Western commentators . . . were highly sceptical of the merits of "typological thinking" in general, and of the approaches of the Polish School in particular; and the Polish participants were, of course, highly critical of each other.'[36]

With the debate centred on such a region at such a time, the Nazis cast a particularly long shadow. Wierciński emphatically rejected the simplistic association between typology and the Nazi race concept.

Hans Günther's typological Nordic supremacism, he argued, had been opposed by the populationist approaches of J. Kaup and Alfred Rosenberg. There might have been a typological basis to Nazi caricatures of Jews, but the National Socialist persecution of Jews was conducted on a populational basis. Jews were exterminated regardless of the colour of their eyes or the shapes of their noses. Robert Proctor makes similar observations, showing that, in Germany, the shift from typological to populational thinking was not accompanied by an abandonment of the belief in racial hierarchy.

Wierciński continued to hold out against the populationist tide right up to the 1980s. The objects of his research included pre-Aztec Olmec skulls in Mexico, among which he detected signs of Caucasoid and Negroid admixture. From this he inferred that both Europeans and Africans had made sufficient contact before Columbus to leave their racial traces among the peoples of Central America. This was precisely the kind of reasoning which made typology look ridiculous in populationist eyes. Wierciński referred to recent work by Polish investigators which had led them to identify traces of the 'Berberic type' in the Congo and in Poland itself. Bielicki pointed out that, thanks to the exuberance of human diversity, it was possible to find individuals in any population who bore a resemblance to a type common in another group, and thus to adduce them as evidence of a historical connection between the two groups. Since humans are inveterately mobile, only the most obviously implausible connections are ruled out. Bielicki claimed that Wierciński or his ilk would be capable of discovering an 'Australoid admixture' among their fellow Poles, on the basis of individuals with the broad noses, brow ridges, long heads and wavy hair commonly found among Australian Aborigines. 'The fact that these individuals would have a light skin and perhaps blue or greenish eyes would not bother the typologist at all: he could simply postulate that "in Cromagnonoid-Australoid hybrids light pigmentation is predominant" – and this would settle the problem.'

By Wierciński's own assessment, the typological (or, as he called it, individualist) concept of race was now upheld only by a minority of anthropologists in Poland, Czechoslovakia and Hungary. This underestimated the extent of typology's adherents, but Central

Europe was certainly its heartland – and remains so, as the Race Gallery in Vienna demonstrates. To the west, the anti-racist line hardened. A few months after the Polish controversy, *Current Anthropology* published a short piece by Frank B. Livingstone entitled 'On the Non-Existence of Human Races', in which he pointed out that biologists trying to classify birds and butterflies had encountered the same difficulties as anthropologists trying to subdivide the human species, and had become disillusioned with the categories of subspecies or race. A race could be defined as a population differing from others in the frequency of one or more genes, but the problem was that the patterns of variation among genes were 'discordant'. Sorting humankind by blood groups or hair type would produce sets of races that differed from each other and from traditional classification schemes. Variation was better described by clines, or gradients, plotted like the isobars on a weather map. 'There are no races, there are only clines,' Livingstone declared.[37]

Theodosius Dobzhansky concluded his critical response to this article with the claim that 'To say that mankind has no races plays into hands [sic] of race bigots'. Anti-racist intellectuals have since tended to draw the opposite conclusion, and Livingstone's clinal manifesto has now been incorporated into radical left criticism of science.[38] If race can be completely denied on scientific grounds, any scientific text using racial categories is all the more suspect. The authority of any biological pronouncement on race is invalidated, and thus the hegemony of the humanities is strengthened. Authority rests solely with those who operate in the cultural domain. The decisive period for the shift of power seems to have been the second half of the 1970s, when the balance of textbook wisdom swung to the position that races do not exist.[39]

It was perhaps inevitable that the race concept would face a challenge like this sooner or later. The mid-century alignment of anthropology and genetics against racism was a compromise between modern and traditional ways of describing human variation. Even the most outspoken anti-racist scientist of that historical moment, Ashley Montagu, was happy to outline a traditional racial taxonomy as late as 1972, although he insisted on the term 'ethnic group'.[40] With its Caucasoids, Nordics and Dinarics, Montagu's classification scheme was similar to that used in the Vienna Race Gallery, which itself cited the 1951 Unesco Statement and its 1962 rewrite as the legitimising authority for its 'value-neutral' racial exhibit. Without values attached, however, the value of the schemes themselves was unclear, and the motives behind them appeared more suspect as time went on. The clinal concept looked like both a scientific rationalisation and a clear-out of political lumber.

Another indicator of the distance travelled in anti-racist politics is Montagu's view, given in the same volume, of why black Americans do worse than whites on IQ tests: 'The Black ghetto child comes not only from a culture of poverty, but from a poverty of culture, parentally uninspired, rootless, barren, and aridly one-dimensional.' In 1992, Frederick Goodwin was forced to step down as chief of the Alcohol, Drug Abuse and Mental Health Administration after observing that 'some of the civilising evolutionary things' had been lost along with social structure in inner cities, so it might not just be 'a careless use of words' to call these zones 'jungles'.[41] In 1972, though, it was okay for Ashley Montagu to call black culture a desert.

Montagu's political perspectives derived from the period before the focus of struggle shifted from civil rights to culture. In the early years of the new Unesco scientific order, the dominant racial issues were integration in the United States, and decolonisation around the world. Among the ranks of the resistance to these historical sea-changes were those for whom a 'value-neutral' race concept was a travesty; for whom all human value was racially marked.

Some of them still kept the cult of the Aryan alive in their hearts, and in 1957 the Northern League was set up to bring them together. In a document setting out its aims and principles, the 'Pan-Nordic Cultural Society' affirmed that human history was essentially racial history, and that outside the Far East, the peoples variously known as Indo-European, Caucasian or Aryan were responsible for almost all the civilisations of the world.[42] However, their influence was disproportionate to their numbers, and they generally formed a small aristocratic elite ruling their subject peoples. Herein lay the seeds of their downfall, since they interbred with these peoples and disappeared. Babylon, Sumeria, Egypt, Greece and Rome all succumbed to miscegenation. According to the document, there was no need to spell out the parallels between the last of these cases, in which the imperial centre was undermined by 'the influx of Roman "citizens" from all parts of the conquered empire', and Britain, which was shedding colonies while admitting immigrants from the former imperial territories. The historical scenario presented by the League was the one previously expounded in *Mein Kampf*, but now the call was to preserve the Nordic biological and cultural heritage in the face of the 'rising tide of colour'. As a living link with the heyday of the tradition, it boasted Hans Günther as a member.

The founder of the Northern League was Roger Pearson, who set out his racial vision in two pamphlets published under an imprint called Folk and Race.[43] In *Blood Groups and Race*, he affirmed that the Nordics were the highest form of life that Nature has ever produced. In *Eugenics and Race*, he urged the revival of racial hygiene, and declared that eugenics and evolution 'constitute the greatest discovery of mankind'. His view of racial struggle was apocalyptic: 'If a nation with a more advanced, more specialised, or in any way superior set of genes mingles with, instead of exterminating, an inferior tribe, then it commits racial suicide, and destroys the work of thousands of years of biological isolation and natural selection.' For Pearson, there was evidently no middle ground, such as separate development, between race suicide and genocide. The future of humanity lay in concerted efforts to preserve 'an aristocracy of mankind' by selective breeding to recreate ideal types. 'We in northern Europe for our part can perhaps hope to recreate a society comparable to that of the Heroes of Asgard.'

The pamphlets, part of a series featuring titles such as *Nordic Twilight* and *This Is Odinism*, were reprinted in 1980, and in the mid-1990s were still being advertised in the mail-order section of *Spearhead*, the magazine of the fascist British National Party. Pearson himself combined an academic career in the United States with political activism. In the mid-1960s he edited a journal called *Western Destiny*, whose contributing editors included Earnest Sevier Cox, a Ku Klux Klan leader; A. K. Chesterton, whose career on the far horizons of the British right took him from the British Union of Fascists through the League of Empire Loyalists to the chairmanship of the National Front; and Arthur Ehrhardt, formerly of the Waffen-

SS and subsequently the founder of a fascist periodical called *Nation Europa*. In 1978, Pearson chaired a conference of the World Anti-Communist League, though he parted company with WACL after a rival faction accused him of neo-Nazi connections. Despite this contretemps, Pearson has succeeded in making friends in respectable right-wing quarters. In 1982, his efforts in the conservative cause – and a complimentary copy of a political journal he publishes – earned him a letter of congratulation from President Reagan.[44]

Pearson also publishes an anthropological journal called the *Mankind Quarterly*, an organ of similar vintage to the Northern League. During the early 1960s, it gave public voice to the kind of ideals the League promoted more privately. It was founded and edited in Edinburgh by Robert Gayre, who amassed a notable collection of scholars to act as Honorary Associate Editors or to sit on the Honorary Advisory Board. The latter category included, among many others, the eminent palaeoanthropologist Henri Vallois, a member of the panel which drew up the 1951 Unesco Statement, the psychologist Hans Eysenck, and the racial hygienist Otmar von Verschuer. Among the journal's early contributors was Ilse Schwidetzky, who described 'Negrid' traits in an article on 'Racial Psychology': 'wide-open moistly shining eyes on the physical side; strong and unstable excitability on the psychic side.'[45]

The association with Vallois notwithstanding, the *Mankind Quarterly* served as a refuge for race scientists who could not accept the Unesco anti-racist order. The most prominent of these was Reginald Ruggles Gates, whose lifelong obduracy on race left him an isolated figure after the war. Gates was the one heavyweight scientist in the English-speaking world who still insisted that, contrary to the opening declaration of the Unesco Statements, there were several species of human being. Two members of the Advisory Board resigned in protest when they actually saw copies of the journal. One was Bozo Škerlj of Ljubljana University, whose role as a contributor to and editorial associate of the *Zeitschrift für Rassenkunde* did not save him from the Dachau concentration camp. Gates, one of the *Mankind Quarterly*'s associate editors, wrote to the Slovenian anthropologist informing him that he would never have been invited to join the Board if Gates had known about this incarceration, 'which naturally had such an effect on' Škerlj's 'mental outlook'.[46]

The *Mankind Quarterly* survived the heavy criticism it received after *Current Anthropology* gave an attack on it the star treatment, but under Gayre's direction it confined itself to the margins of the political right.[47] Gayre's preoccupation with 'separate development' in Southern Africa led him to run reviews of publications such as the 1965 Rhodesian Ministry of Internal Affairs Annual. Closer to home, in 1969 he recommended an edition of Hans Günther's *The Religious Attitudes of the Indo-Europeans*, issued by the Britons Publishing Company, among whose other titles was the notorious anti-semitic forgery *The Protocols of the Learned Elders of Zion*.[48]

By the mid-1970s, Gayre felt able to find space for claims that the existence of the Yeti had been proved.[49] Many of the names in the editorial lists were now marked with obituary crosses, such as those of Gates and von Verschuer, but the journal also acquired a notable new recruit, the British psychologist Richard Lynn. In 1978, when control of the *Mankind Quarterly* passed across the Atlantic from Gayre to Roger Pearson, Lynn remained an associate editor. Under Pearson's control, the publication conforms much more closely to academic convention, but its racial disposition continues to exclude it from the mainstream. It remains an organ for what is beyond the pale elsewhere. While Richard Lynn has other outlets where he can discuss the racial implications of differences in IQ scores, for example, the *Mankind Quarterly* is the place where he can air his views on the levels of civilisation attained by different races.[50]

The *Mankind Quarterly* acquired a German cousin in 1972. *Neue Anthropologie* was published by the *Gesellschaft für Biologische Anthropologie, Eugenik, und Verhaltensforschung* (Society for Biological Anthropology, Eugenics and Behavioural Research), under the editorship of Jürgen Rieger. Its first issue honoured the former racial hygienist Fritz Lenz on the occasion of his eighty-fifth birthday. Rieger had previously paid tribute to Hans Günther at meetings of the Northern League, where he urged his fellow Nordics to 'fight for a Teutonic confederation'. In 1977, he helped organise a meeting on 'Eternal Penitence for Hitler?', featuring Arnold Butz, who specialises in denying the extermination of the Jews, and Gerhard Frey, the chairman of the ultra-right *Deutschen Volksunion*. Several active neo-Nazis were members of the Board of Scientific Advisors, as was the founder of the hereditarian IQ school, Arthur Jensen.[51]

As if to underscore the point that the world of hardline race science is a small and tightly knit one, a minor scandal erupted within the German delegation to the World Population Conference, held in Cairo in 1994. Charlotte Höhn, director of the Bundesinstitut für Bevölkerungsforschung (Federal Institute for Population Research), or BiB [sic], gave an interview to the left-leaning Berlin newspaper *die tageszeitung*, in the course of which she remarked that the average intelligence of Africans is lower than that of others, and questioned whether sick people should be allowed to have children. This seemed a reasonably faithful interpretation of the BiB's stated mission, to work on 'problems of falling birth rates, European migration and collaboration with developing countries on population issues', given the history of the institute and the man who defined these goals, Hans William Jürgens of the University of Kiel.

The BiB is an instance of what might be dubbed the Strangelove Syndrome, though with *Rassenkunde* rather than rockets. After the war, with their discipline under a cloud, German population scientists regrouped in two private institutes. Their persistence was eventually rewarded in 1973, with the establishment of the public BiB. Jürgens, its first director, made his academic debut in 1961 with a thesis on the biology of the 'asocial' that ended with a call for their sterilisation. It should by now come as no surprise that Jürgens is an associate editor of the *Mankind Quarterly*.[52]

Charlotte Höhn was forced to resign as a result of her indiscretion, but in general the post-war history of conservative race science is one of successful adjustment to reduced circumstances. Writing in the mid-1980s, Benno Müller-Hill argued that the old guard, the Strangeloves, actually guaranteed that a new National Socialism would not arise. They were forced to remain silent in order to retain their chairs, but their students may not feel the same inhibitions. 'I suspect that, with the retirement of these academic cripples of the older generation, the period of tolerance in West Germany is coming to an end,' Müller-Hill observed.[53]

Charlotte Höhn and Hans Jürgens illustrate the ability of the old race science tradition to reproduce itself, and of the second generation to consolidate its institutional base. The network built around the *Mankind Quarterly* and kindred enterprises amounts to a race science international. But the air of intrigue and the lurking swastika should not be allowed to define racial science at the end of the century as a mere cult activity. The prime site for an inquiry into racial questions is the main body of science itself.

In *The Language of the Genes*, the British geneticist Steve Jones recalls that when he read Ashley Montagu as a schoolboy he found Montagu's arguments 'unconvincing and hard to follow'. 'Rereading it recently showed why: Ashley Montagu had tried, nobly, to make bricks without straw,' Jones continues. 'The information needed to understand our own evolution was simply not available at the time and there seemed little prospect that it ever would be.'[54]

Now, Jones says, the situation has been transformed. Vast amounts of data about the human genome and its geographical variation have been gathered; much more is on the way. 'At last there is a real understanding of race, and the ancient and disreputable idea that the peoples of the world are divided into biologically distinct units has gone for ever.'

Yet Johann Friedrich Blumenbach recognised that human variation was continuous, and so did Darwin, and so did the scientists who established the anti-racist order after the Second World War. There has certainly been an enormous increase in the quantity of data supporting the anti-racist position. The portrait of human variety is now coloured and shaded and embellished with a myriad intricate details, but the lines of the sketch were already drawn half a century ago.

In those days anti-racist science was combative. It has subsequently become anodyne, smoothing over the race issue with the bland reassurance that modern population biology is intrinsically antithetical to racism, and that race science is confined to a small coterie with dubious motives. It is often denied that this faction is part of science at all. As Donna Haraway remarks, 'pseudo-science' is 'always defined so as to show how current science escapes any such taint'.[55] It just isn't that simple. Race remains embedded in science at many levels, as do the hardline race scientists themselves.

3

Antipodes

THERE IS AN odd man out in the Vienna Race Gallery. He is sombre, not smiling; his eyes are fixed on an uncertain horizon, and he is not a photographic image but a painting. There is good reason for his prominent nose to be out of joint, because he is a Tasmanian. The catalogue singles out his people as the first variety of human to have become extinct in recent times.

According to the gallery's account, relations between the white settlers and the Tasmanian Aborigines were relatively good, but the colonists encroached on the nomadic hunter-gatherers' *Lebensraum*. The indigenes were assembled by a missionary, George Robinson, and placed in a settlement on Flinders Island. The measure was well-intended, but its effects disastrous. By 1847 only 47 of some 2,000 Tasmanians were left. The last full-blooded Tasmanian, Truganini, died in 1876.

A rather different story unfolds in another account of human biology aimed at a general audience, Jared Diamond's *The Rise and Fall of the Third Chimpanzee*.[1] Diamond presents his account as 'a case study typical of a broad class of genocide'. He gives the population at the time of contact as 5,000 rather than 2,000; their material culture did not include pots, nets, or the ability to light fires. Diamond uses the same term as the Vienna catalogue, *Lebensraum*, to describe the ground of the conflict between the Tasmanians and the British who began to arrive around the turn of the nineteenth century. Unlike the catalogue, however, he describes the British tactic of kidnapping – children for labour, women for sex – combined with the killing of men. After reprisals, natives were banned from areas settled by whites, and killed by 'roving parties' in which the officers were police and the men were convicts. Subsequent measures included martial law, under which soldiers were authorised to shoot Tasmanians on sight, and bounties – £5 for an adult, £2 for a child.

Diamond agrees that George Robinson believed he was acting in the natives' best interests. But he adds that some of Robinson's charges were rounded up at gunpoint, that children were separated from their parents at the Flinders Island settlement, that it was run like a jail, that the diet caused malnutrition, and that the government cut the funding in order to hasten the inmates' end.

Apart from the discrepancy over numbers, the two accounts are mutually consistent. They even use the same German word and share an assessment of individual motivation. Yet they are of course utterly different. The museum catalogue attempts to maintain the gallery's value-neutral facade by excluding historical testimony that clearly indicates a genocidal policy on the part of the Europeans, while putting in a word for Robinson's good intentions. The Race Gallery survives under the post-war anti-racist order by adhering to the letter of the later Unesco Statements while ignoring the humanist spirit behind the original Unesco project. Diamond, by contrast, discusses the Tasmanian episode as part of an inquiry into the biological dimensions of genocide. This is precisely the kind of popular and political engagement by scientists that Unesco had ordained when it identified racism as a cause of the Second World War, and made science the foundation of anti-racism. The Vienna gallery sticks to the new rules, but is deeply unwilling to acknowledge the historical reality of genocide, whether on the other side of the world, or closer to home.

Here, 'value-neutral' human science means detachment from human values. Having edited out the case for genocide charges, but described the extinction of the Tasmanians, the gallery text announces that 'by good fortune' the museum possesses a Tasmanian skull. The scientific material thus survives, and one is left to wonder whether, for 'value-neutral' science, that is what counts. The value of the material itself is conveyed by Diamond's account of the division of the anatomical spoils from the last Tasmanian man. Between the competing Royal College of Surgeons and the Royal Society of

Tasmania, tussling over the remains, an *ad hoc* Solomonic solution emerged. Dr Crowther of the Royal College got the head, though not the ears and nose, which went elsewhere; Dr Stokell of the Royal Society made off with the hands and feet, and also the skin, from which he fashioned a tobacco pouch.

The last survivor, Truganini, pleaded to be buried at sea, so as to avoid a similar fate. In the event she was buried on land and later exhumed; her skeleton was exhibited in the Tasmanian Museum until 1947, when it was taken off public display and reserved for the scientific gaze. Nowadays all such scientific material is the object of contests not between rival scientists, but between curators and indigenous peoples' political activists; between Western scholarly values and non-Western spiritual ones. Truganini's posthumous supporters eventually achieved a sort of victory: in 1976, the centenary of her death was commemorated by the cremation of her bones and the scattering of her ashes at sea, a belated approximation to her own wishes.

The Tasmanian lineage has not died out altogether. Though the scientifically prized full-bloods were gone with Truganini, Tasmanian genes persist in a number of people descended from matings between intruders and natives. They defy both the genocidal project and the assertions of nineteenth-century race scientists that Europeans and Tasmanians were biologically too far apart to produce fertile offspring. 'Mulism', as it was known, was considered to be the normal outcome of hybridisation between humans of widely separated stock. Its occurrence could be relied upon in the Antipodes, the opposite ends of the Earth, since the peoples of the Australian continent and its periphery were taken to be the antipodes of the European races: the opposite extreme to the Europeans' supposed physical and cultural advancement. In certain respects, that assumption endures.

For traditional morphologists, the Australian skull is an object of wonder and of exceptional reassurance. It bears the stigmata of primitiveness; no fewer than twenty-eight of them, according to a zoologist named John R. Baker, whose book *Race* appeared in 1974. His perspective is expressed in his opening remarks about the 'Australids'. A group of individuals that become isolated, he observes, may 'fail to evolve further, or make only limited advances'. Such a population 'may serve as a reminder of a stage in the evolution of more advanced forms'.[2]

Baker's treatise, compendious and ponderous, is possibly the last major statement of traditional race science written in English. It can also claim the distinction of being almost certainly the only book to be endorsed by both the eminent scientist Sir Peter Medawar and the fascist British National Party. On its jacket, Sir Peter praises the 'thoroughness, seriousness and honesty' of the work. 'The idea of race or raciality has been systematically depreciated for political or genuinely humanitarian reasons,' he continues, 'and it was high time that someone wrote about race as Baker does, i.e. in the spirit and style of a one-man Royal Commission.' The listing in the Book Service section of the BNP magazine *Spearhead* commends it as 'an objective and scholarly account' of 'great importance'.

Baker's central purpose is to re-state the case for racial hierarchy, to which end he draws a comparison with seagulls. A birdwatcher describing a circle around the North Pole in a westward direction would find the herring gull, *Larus argentatus*, growing steadily larger, its back and wings getting darker, and its feet turning from pink to yellow. The populations that would be encountered on this journey can be divided into seven races, or they can be considered as a single cline. In the White Sea off north-western Russia, however, two of these races mix with each other but fail to interbreed, which would ordinarily lead to their classification as separate species. Meanwhile, a birdwatcher in a British harbour might well be able to tick off both herring gulls and lesser black-backed gulls, the latter being assigned the species name *Larus fuscus*. Yet lesser black-backs, with their dark wings and yellow feet, look like the final stage in the westward trend of herring gulls. One variety of herring gull has been reported to interbreed with the British variety of black-backed gull. And so on, through a total of seventeen putative races; under the circumstances, Baker concludes, the idea of species is best dropped altogether, in favour of a group of interrelated races which shade into each other without necessarily interbreeding.

In other domains of scholarship, this is called deconstruction. Baker's aim is to undermine the opening assertion of the Unesco

Statements, that all humankind belongs to a single species. Instead of trying to defend an untenably strong argument to the contrary, like Ruggles Gates, he denies the importance of the species and emphasises that of variation. He also suggests that human nature is itself unnatural, creating conditions which erode natural barriers to interbreeding. This applies not only to humans themselves, but to other creatures under their control; thus herring gulls and common gulls do not interbreed in the wild, but do so in captivity.

The extent to which this happens is a function of domestication: taking Blumenbach's insight that humans are 'of all living beings the most domesticated', Baker argues domestication has attenuated humans' natural revulsion against interbreeding so far that bestiality must be prevented by law. Having cast doubt on the concept of the species in general, Baker concludes that, even if the idea has validity in some biological circumstances, 'it does not appear to be applicable to human beings'.

Unlike the more familiar sort of deconstructionist, however, Baker retains an unshakable belief in progress. There is no doubt in his mind that evolution is going somewhere, that an overarching trajectory exists for Man to follow, onward and upward, along which an isolated population like that of Australia may fail to progress. This belief persists in newer evolutionary theories of race, which implicitly use ethnic competition as their criterion of biological fitness. One way of conceiving the fitness of a population, such as that of Australia or Tasmania, is to consider such factors as the ability to make use of diverse food resources, to cope with climatic variations, and to conserve energy. Another is to consider what happens when another population expands into its territory. When the intrusive population is a technologically advanced one from Europe, the reckoning is unarguable. You can't stand in the way of progress.

The language of cultural relativism does seem to reach the limits of its credibility in the technological sphere. Machines, unlike biological structures, are purpose-built. A spear and a rifle are both artifacts designed for killing; the latter is more advanced than the former, and has been adopted by traditional hunters from the Arctic to Australia. But the choice of example is loaded. Donna Haraway argues that

Man the Hunter, in his modern form, was created as a foundation for Unesco Man, the Universal Man rediscovered in the aftermath of world war.[3] Man was conceived as the result of a limited number of defining adaptations. The first of these, for which the fossil australopithecines furnished the evidence, was bipedalism. First the ape stood up, freeing its hands and allowing it to use tools. In doing so, it began a dialectic of culture and nature which drove hominisation. The upright ape, *Australopithecus*, became Man, *Homo*. As the brain expanded in the course of becoming human, the teeth became smaller. Hominids lost their built-in weapons, but developed the possibility of building their own, all the way to the Bomb. Among females, the complementary development was the loss of oestrus and thus the removal of time limits on sexual 'receptivity': the more effective bonding this permitted between mates was an enabling condition for the human family.

Haraway points out that in these ways, the distinction between the sexes was reduced; she argues that the same 'narrative logic' removed meaningful racial differences from the story of human evolution. In the new account, human bodies were universal, unmarked by the differentiating characteristics of sex and race. But the final adaptation, which defined the genus *Homo*, was masculine. This was the extension of the hominid diet by means of hunting, an activity which developed the human co-operative spirit, in males, at the expense of other animal species. The principal author of this vision, Sherwood Washburn, insisted that the hunting way of life – not the farming one – was the basis of human unity, human universals, human mental and behavioural plasticity. This founding myth could not survive the critical scrutiny of feminist anthropologists in the 1970s and 1980s, when universals got a bad name – imperialism, racism, and so on. Nevertheless, although Washburn's Man the Hunter may have been fatally flawed by masculinism, Haraway draws attention to his role at a particular historical moment in the cause of racial equality and humanism. 'The contradictory creature produced through the hominizing behavior of hunting was a natural global citizen,' she comments, 'as well as a natural neo-imperialist; a natural political man, as well as a natural sadist; a natural providential father and reliable colleague, as well as a natural male supremacist. His plasticity defined him; his most fundamental pleasure threatened him with extinction.'

The antipodes of Man the Hunter, one wielding spears and the other rockets, co-existed on the Australian continent during the Cold War. As Britain was shedding its imperial territories, it developed the nuclear weapons that would guarantee the continuation of its global influence. It tested the Bomb on the Monte Bello coastal islands, in the Woomera rocket range, named after an Aboriginal word for a spear-throwing device, and in the adjoining Maralinga area. During the same period, the American anthropologist Carleton Coon conducted his own researches on islands off the Australian coast, on an Aboriginal people called the Tiwi. They lived on Melville and Bathurst Islands, a hundred kilometres or so north of the regional capital of the Northern Territory, which is called Darwin.

For Coon, the cultural contrast between European and Aboriginal Australian spoke for itself. A spear thrower and a rocket launch-pad could only be connected by a metaphor of function. The concept of mental plasticity could not be stretched to cover the gap between the two peoples. Coon took it as evidence of polygeny. 'If all races had a recent common origin, how does it happen that some peoples, like the Tasmanians and many of the Australian aborigines, were still living during the nineteenth century in a manner comparable to that of Europeans of over 100,000 years ago?' he asked. 'Either the common ancestors of Tasmanians cum Australians and that of the Europeans parted company in remote Pleistocene antiquity, or else the Australians and Tasmanians have done some rapid cultural backsliding, which archaeological evidence disproves.'[4]

Coon emphasised the persistence of technologies which were primitive even by Stone Age standards. Up until a couple of generations before, the Tiwi had used crude chopping tools of a kind which first appeared half a million years ago; some of their cousins possibly still did. 'Tiwi society is undeniably archaic,' Coon stated. 'The Tiwi lie on the fringe of a marginal continent; they are the most marginal of marginals. They have never had spear throwers, stone-tipped spears, boomerangs, circumcision, or other elements of "advanced" Australian aborigine culture.'

He detected a similar archaism in the skulls of the Tiwi, as in those of Australian Aborigines as a whole; primitive biology, he believed, determined primitive culture. The characteristic features of Australian skulls were also those of the primitive modern human condition: prognathism, heavy brow ridges, and a thick, flat cranium similar to that of *Homo erectus*, who was abroad in the Australasian region at an earlier stage in prehistory. In *erectus*, this 'looks somewhat like a poorly raised loaf of bread'; in modern Australians, it is called the 'ill-filled' look. John R. Baker also drew attention to this feature, describing it as reminiscent not only of *erectus* (or *Pithecanthropus*, his preferred term) but of pongids, or apes.

Recently, researchers have suggested that the remarkable thickness of many Australian skulls is the result of an unusually literal sort of selection pressure, a ritualised tradition of settling disputes by a series of blows to the head with a digging stick; from fossil skulls bearing depressed fractures, the practice is inferred to be at least eleven thousand years old.[5] Peter Brown, of the University of New England in New South Wales, suggests that injuries today may be more severe, partly because more dangerous instruments are being used, and partly because interbreeding with whites has resulted in thinner skulls.

Coon used a version of the facial angle, the mandible angle of inclination, to recreate a Great Chain of Being in which superiority was marked by increasing steepness. The angle in the orang-utan was 44°, in australopithecines 58°, in *Sinanthropus* (a Chinese variety of *Pithecanthropus*) 61°; in Australians it was 75°, and reached the perpendicular in Whites. He offered a parallel hierarchy based on cranial capacity: the average volume of the male *Pithecanthropus* braincase was 900 cubic centimetres; that of 'Solo man', late *erectus* remains found in Java, was 1,150cm^3, and that of living Aboriginal Australian men was 1,350cm^3. The ascent of woman took place on a lower level, from 775cm^3 in *Pithecanthropus* through 1,040cm^3 in Solo females, through to 1,180cm^3 in modern Aboriginal Australian women.

The fact that these were average figures implied a degree of overlap in the range, so that some modern Australian women had brains smaller than those of some male *Pithecanthropi*. 'Some evolution has probably been taking place,' Coon allowed, 'but, as one would expect in a marginal area of the Southern Hemisphere, its over-all rate cannot have been rapid. One still finds recent aboriginal female skulls with cranial capacities of 930cc., 946cc., and 956cc. whose owners apparently met the demands of their culture well enough to live to maturity.'

He illustrated his point with a photograph of one such woman, a Tiwi named Topsy, whom he calculated to have a cranial capacity of less than a litre. Her picture was placed next to that of 'a Chinese sage with a brain nearly twice that size (Dr. Li Chi, the renowned archaeologist and director of Academia Sinica)'. The antipodean pair were labelled 'The Alpha and Omega of Homo sapiens'.

This plate became one of the principal exhibits supporting charges of racism that followed the publication of *The Origin of Races* in 1962. Coon found himself marginalised, his theories considered both unpalatable and implausible. Recently, however, several popular books – Erik Trinkaus and Pat Shipman's *The Neandertals*, Pat Shipman's own *The Evolution of Racism* and Christopher Wills's *The Runaway Brain* – have taken it upon themselves to mount a posthumous defence on his behalf.[6] They may be taken as straws in the wind.

Coon strained credulity because he proposed an evolutionary convergence between independent hominid populations, towards a common *sapiens* state. His theory was based on the work of Franz Weidenreich, a Jew hounded out of his academic post by the Nazis, who went on to study what are now classed as *Homo erectus* fossils in China. Weidenreich considered that these bore Mongoloid racial traits, though many of his peers believed that racial differentiation took place much later. According to Weidenreich, races were hundreds rather than tens of thousands of years old; older even than the modern human species. He suggested that humanity evolved as a global system within which genes were exchanged, permitting species unity, but which also preserved regional characteristics.

Coon's version played down the gene flow and emphasised the differences between populations. He argued that *Homo erectus* had spread around the Old World, then divided into five subspecies. Each of these had then evolved, of its own accord, into *Homo sapiens*.

The result was a single species composed of five races: Caucasoid, Mongoloid, Congoid (equivalent to Negroid), Capoid (represented today by the Khoisan of Southern Africa) and Australoid. Critics like Theodosius Dobzhansky considered the possibility that *Homo sapiens* arose not once but five times to be 'vanishingly small'.[7]

Coon's vision of human origins is convincing enough if you have a profound belief in the progressive nature of evolution. The idea of 'grade' was fundamental to his thinking: the *sapiens* grade was a stage of evolutionary advance reached at different times as different human subspecies 'passed a critical threshold from a more brutal to a more sapient state'. The subspecies which made it first have evolved the most, and this is reflected in the degree of civilisation they have attained, as well as in morphology. It was clear from their physical characteristics that 'Australian aborigines are still in the act of sloughing off some of the genetic traits which distinguish Homo erectus from Homo sapiens.'

Small-brained Australian Aboriginal women like Topsy were hostages to his theory. If fossil skulls of similar 'grade', like those found at Zhoukoudian near Beijing, were not *sapiens*, then neither were Topsy and her sisters. By extending the *sapiens* category, the gap between modern humans and ancient ones such as the Neanderthals was closed. Coon agreed that there were clear differences between fossil specimens and 'the articulated skeletons of urban paupers dangling from hooks in European and American lecture rooms'. But the picture changed if one looked towards the marginal populations such as those of Melville Island, where 'brow ridges reach their peak'.

Melville Island did not become Brow Ridge Capital of the World just because it was a long way from anywhere that mattered. The Tiwis' technological level, their degree of civilisation, and their state of physical evolution were all determined by evolutionary selection pressure, or rather the lack of it. Coon summarised their condition thus: 'Physically they are . . . archaic full-sized human beings with a plethora of heavy brow ridges and big teeth, and brains of only moderate size. They have had the fortune to be preserved in a geographical paradise in which an early and agreeable form of human life can be led by healthy people without too much effort . . .' Similarly, the evolutionary engine only got into higher gear when humans left the benign, unchallenging environment of Africa for the harsher climes of Asia and Europe. 'If Africa was the cradle of mankind, it was only an indifferent kindergarten,' Coon concluded. 'Europe and Asia were our principal schools.'

In Coon's vision, adaptation was everything. If the assembly of characteristics that constituted *sapiens* was to any great extent the result of chance, its emergence five times would indeed be incredible. It only makes sense if the traits of modern humanity are the result of Darwinian forces that act in all environments. This is akin to the *Star Trek* convention that the default condition for intelligent life throughout the Universe is humanoid. As well as drastically reducing the requirement for animation, it expresses the idea that the condition of humanity is somehow predestined.

The antipode to this concept of evolution, as expressed in the writings of Carleton Coon and John R. Baker, is the vision articulated by Stephen Jay Gould in his book *Wonderful Life*.[8] Gould describes an assemblage of fossils found in a geological tract called the Burgess Shale. A remarkably high proportion of them represent types of organism which have vanished, yet there is no apparent reason why some should have survived and others have become extinct. One form seems pretty much as good as another. Survival seems to have been more a matter of chance than of adaptation, which makes us a product of contingency rather than inevitability. Some readers seemed to find this challenge to the idea of manifest human destiny highly unpalatable. Perhaps attachment to the idea of manifest white male destiny also runs deeper than contemporary rules of public discourse might suggest.

The theoretical current of which Coon and Baker were part did not dry up with them. Now expressed in what is known as the 'multiregional hypothesis' of modern human origins, it lives on as a vigorous challenger to the school of thought which proposes that modern humans all share a common African ancestor. And race is the spectre that haunts the debating chamber.

The argument has become personalised around the leading advocates of each theory; Milford Wolpoff of the University of Michigan for multiregionalism, and Chris Stringer of the Natural History Museum in London for the 'out-of-Africa' proposal.

Wolpoff and his followers have returned to Weidenreich's original concept, restoring the gene flow between different populations without which Coon's model lacked credibility. Stringer argues that while earlier human varieties spread throughout the Old World, they were replaced by a modern population which expanded out of Africa, but did not interbreed to any significant extent, if at all, with the populations it encountered.

Wolpoff and Stringer are both physical palaeoanthropologists, the core of whose expertise lies in the study of skulls, and they each believe that the fossil evidence supports their own positions. Flexing its muscles as the new paradigm for all of human science, genetics has offered itself as a means of breaking the deadlock. A triumphalist claim on these lines was made as far back as 1971 by the outspoken molecular anthropologist Vincent Sarich, who declared that it didn't matter what the fossil evidence looked like, since the answers would be found in the molecules.[9] It was not until the mid-1980s, however, that genetics made a really dramatic claim for explanatory power over modern human origins.

The Eve hypothesis emerged with the publication in 1987 of a paper by Rebecca Cann, Mark Stoneking and Allan Wilson, in which they described a survey of mitochondrial DNA from around the world.[10] Mitochondrial DNA exists in structures outside the nucleus of a cell, and so is inherited only via the egg, not the structurally rudimentary sperm. It therefore forms a record of maternal descent. As it was found to be most variable in Africans, the researchers concluded that this was where the lineage was oldest. By using a computer program to construct a 'tree' of relatedness, they worked out that all existing humans were descended from a woman who lived in Africa about 200,000 years ago. Although this was not the same as saying that she was the only woman alive at the time, she inevitably became known as Eve.

Her attractions were many and powerful. As a scientific concept, she was easy to grasp, and easy to illustrate in magazines. She restated, personified and updated the Unesco opening proposition, which with a woman at its head now declared that all humankind was one. At the same time, she reconnected science with the Biblical tradition of monogeny from which it had pulled away in the first half of the nineteenth century.

As it turned out, she was too good to be true. Critical evaluation of the original paper revealed shortcomings in the procedure. It was shown that the shape of the tree could be affected by the order in which the data were entered, and that thousands of trees could be generated, not all of them rooted in Africa. More trenchantly still, the geneticist Alan Templeton challenged the equation of mitochondrial DNA patterns with actual human populations. His own analysis inclined him towards the kind of gene-flow model proposed by the multiregionalists.[11] Each camp now has its geneticists.

Whatever the future course of the debate initiated around Eve – and it has certainly not yet run its course – the episode helps to illustrate the impossibility of making simple associations between theories, disciplines and race. The emphasis multiregionalism places on primordial racial traits, older than modern humanity, might seem to make it the more attractive hypothesis for those who still believe in racial hierarchies. Yet, as will become clear further on in this story, the out-of-Africa hypothesis tends to be the choice of the new generation of evolutionary race theorists.

There is also no neat alignment between older and newer schools of thought, with physical anthropology, craniology, typology and race on one side; genetics, populations and clines on the other. There is an argument that palaeoanthropology is bound to remain typological, since it has no populations to study, and has to rely on a meagre and fragmentary accumulation of fossil remains. This is rejected by C. Loring Brace, a colleague of Wolpoff's at the University of Michigan. For thirty years he has argued that palaeoanthropology is capable of taking evolution on board, but has largely failed to do so. He makes an exception for the Wolpoff school, while accusing Chris Stringer of operating on principles 'rooted in the pre-Darwinian traditions of neo-Platonic essentialism favoured by Medieval scholasticism'.[12]

The positions taken by Stringer and other out-of-Africanists are not as hard-line as their opponents make out, and if there is one area in which they are particularly soft, it is Australasia. In their original paper, Cann, Stoneking and Wilson proposed that modern humans replaced *Homo erectus* in Asia without 'much' mixing, but they did not absolutely exclude the possibility. Stringer agrees that the evidence for regional continuity is best in Australasia, and has always considered it a possibility there; though he still considers Africa to be the most likely source of the large cheekbones and strong jaw muscles of the Australian face.[13]

It seems difficult, if not impossible, for palaeoanthropologists to view Australian Aborigines as other than archaic. The technical term is plesiomorphy, which means primitiveness of form. If the assumption of mental plasticity is correct, then Australian traits are no more than a matter of form, with no implications for mental capacity. In this view, the great leap forward to modern humanity established a level of mental development that formed a constant throughout all modern human populations. If Australians continue to look more like early modern humans than other populations, this is probably only a historical quirk, the result of their early arrival in the continent and subsequent isolation.

Attempts to test Australian Aboriginal intelligence have had predictable results. The pioneer in the field was Stanley D. Porteus, who spent half a century on his investigations into what he called 'primitive intelligence', beginning in 1915. His favoured instrument was his Maze test, which he believed overcame the cultural differences between subjects and investigators. Aborigines always performed worse than whites, but the more contact they had with white Australians, the more their scores improved. The investigators McElwain and Kearney used a different sort of 'culture-free' test, but found that in places where blacks and whites had mixed extensively, the two groups achieved similar scores. They eventually lost faith in the possibility of devising a genuinely culture-free test.[14] Porteus remained satisfied of the validity of his Maze and of Aboriginal inferiority, however, offering his reminiscences of 'chasing primitives over large continent sections' as a contribution to the *Current Anthropology* debate on race and to the *Mankind Quarterly*, of whose Honorary Advisory Board he was a member.

His compatriot A. P. Elkin remembered challenging his views, but reflected regretfully upon the failure of any 'full-blood' Aborigine to adopt 'our civilisation', the only Aborigines playing 'a responsible role as citizens' being of mixed descent.[15] In the late 1940s, Elkin

recalled, he had contributed to discussions surrounding the preparation of the United Nations Charter. (Around that time, incidentally,
he had also advised the Australian government to ignore arguments
that the planned Woomera rocket range would encroach on
Aboriginal reserves.)[16] The inclusion of Australian Aborigines in the
Declaration of Human Rights was an 'act of faith and hope' that they
and other such peoples could 'appreciate and benefit from those
rights, adapt themselves to modern civilisation, and take their places
in the world's political and economic system'. So it was not a
Universal Declaration, but a contract conditional upon acceptance of
the globally dominant culture. Certain human groups might lack the
mental capacity to qualify for human rights after all. Elkin did not
spell it out, but this would have justified the Australian government's
denial of civic rights to Aborigines, who were not granted the vote
until 1967.

The discourse on the human biology of Australians is muted by
the latent threat of a connection between the idea of plesiomorphy,
even defined as quite superficial, and this kind of full-blooded
racism. In one area, however, it is considered safe to acknowledge
the existence of distinctive biological traits: that of climatic adaptation. As noted earlier, and thanks in large part to Carleton Coon,
climate serves as an acceptable explanation for how northern peoples
got their narrow noses. Australian Aborigines have flat noses and
long thin legs, like most Africans, and likewise live in predominantly
warm conditions. In the central desert, however, they encounter
cold nights; they respond with a narrowing of their blood vessels
which cuts blood flow to the limbs, enabling them to maintain their
basal metabolic rate in temperatures that would force the metabolic
rates of Europeans to rise by 15 per cent.[17]

Yet even this feature, interesting but apparently innocuous, can be
interpreted as a sign of inferiority. The *Cambridge Encyclopedia of
Human Evolution* observes that 'Vasoconstriction in central Australian
Aborigines ... allows them to exploit a savanna with cold nights
without the need for clothing, housing or extra food.' Although
barely visible in a work that projects explicitly liberal messages on
race, the implication is there: counter to the general principle that
humans use culture to adapt to their environments, Aborigines have a

biological adaptation to compensate for the primitiveness of their material culture.

A similar interpretation lurks behind the observation that Aborigines have the largest teeth of any living people.[18] Teeth get smaller, the argument runs, when technology becomes more advanced. We are back to Thomas Huxley's prognathous relative, who depends on bites more than thoughts. Yet Loring Brace has used it in an argument which he claimed 'should finally and permanently dispel the widespread assumption that the Australians were a primitive vestige of Pleistocene human form'. His own analyses of Australian teeth led him to conclude that they were larger in more benign regions, and smaller in harsher ones, such as the central desert. He reasoned that advances in food preparation techniques had permitted tooth reduction, along with other improvements in the utilisation of resources which had enabled people to move into the inhospitable parts of the continent. Such improvements, he considered, had been introduced into Australia from elsewhere. Brace's target is typology, and in this case the 'myth' of a homogeneous Australian type, rather than the general idea of Australian primitiveness.[19]

The persistence of this notion is an indicator of the extent to which Western science continues to assess culture as a whole in terms of material culture. While physical anthropology remains preoccupied with physical objects, however, cultural anthropology has identified a domain of mental life in which the Australian Aborigines can be hailed as world champions. In 'Race and History', written for Unesco in 1952, Claude Lévi-Strauss observed that

In all matters touching on the organization of the family and the achievement of harmonious relations between the family group and the social group, the Australian aborigines, though backward in the economic sphere, are so far ahead of the rest of mankind that, to understand the careful and deliberate systems of rules they have elaborated, we have to use all the refinements of modern mathematics . . . it is no exaggeration to say that they are not merely the founders of general sociology as a whole, but are the real innovators of measurement in the social sciences.[20]

While Lévi-Strauss saluted the Aborigines as fellow structuralists, other whites have based their admiration upon different qualities they perceive in the First Australians. The doubts of Stanley Porteus and A. P. Elkin about the ability of Aborigines to participate in Western civilisation are not shared by liberal whites. From *Crocodile Dundee* to Bruce Chatwin's *The Songlines*, the Aborigine is now commonly depicted as having a shrewd ability to understand white culture – while taking advantage of white prejudice by feigning cultural incomprehension.

In the eyes of whites beset with anxiety about the values and possibilities of technological civilisation, the ability to move between two worlds invests the Aborigine with the mythic powers of a shaman. It is more than a matter of admiration for his cognitive skills in the wild or his mysterious symbolic universe. He – this archetypal Aborigine is male – is felt to be in touch with fundamental truths about human existence that have been lost as material culture has been developed. This has something to do with Christian and other religious traditions that regard austerity as essential for spiritual development.

It also has a close relationship to Sherwood Washburn's vision of Man the Hunter, which is one reason for the masculinity of the archetype. Trying to build a model for a new Universal Man in the shadow of the Bomb, Washburn was haunted by the possibility that breaking free from natural constraints was a false liberation that would end in disaster. 'Man meets the problems of the atomic age with the biology of hunter-gatherers,' he warned.[21] The Aborigines watching the mushroom clouds rising over Maralinga and the scientists who made the bombs were equal after all, and therein lay the problem.

A similar pessimism descended on Carleton Coon towards the end of his life. In the final sentences of one of his last books, *The Hunting Peoples*, he proclaims the need to 'learn how nature intended human beings to live', and, in an opaque formulation, to 're-establish continuity with those who may still be alive after the rest of us are dead'. These are precisely the peoples whom he had previously dismissed as marginal, or even the marginal of the marginal. Otherwise, he concludes, '*someday, out in the desert, a few families of hunters may meet, and ask one another: "Where has whitefella gone?"* [22] (Italics in the original.)

Coon's preceding remarks evoke the dissatisfaction with consumer society, and the social relations it engenders, that was expressed in the 'counter-culture' of the period in which the book was written. It follows the logic of evolutionary competition through to its conclusion: if the apparent victors of the ethnic struggle go on to wipe themselves out, then technological superiority is actually a massive Darwinian handicap; and better, then, to be able to narrow one's blood vessels than to rely on manufactured clothes. At the time, however, most conservative racial theorists were more concerned with struggle under actually existing capitalism; and they are all the more so nowadays.

4

Can White Men Jump?

AUSTRALIAN ABORIGINES ARE some of the leggiest people in the world. In Koreans, mean sitting height amounts to 54.1 per cent of standing height; in African-Americans, it is 50.8 per cent, and in a sample of 205 Australian Aboriginal men, it was 47.7 per cent.[1] This is not surprising, since it conforms with Allen's Rule, which states that 'in warm-blooded species, the relative size of exposed portions of the body decreases with the decrease of mean temperature'. Thus the Inuit of the Arctic are stocky and short-limbed; conversely, people who live in hot climates tend to have longer limbs.

The phenomenon is particularly noticeable in the northern and eastern regions of Africa, where slender people with long legs are common. The anthropologist Jean Hiernaux dubbed this type the 'elongated African'. Rather than being the product of mixture between racial stocks, as the typological tradition supposed, he considered it to be an adaptation to hot dry conditions; the evolutionary interpretation is widely accepted.

For contemporary European aesthetic sensibilities, the elongation of the African often results in exceptional beauty. As well as honouring the ideals of tallness and slenderness, it is often accompanied by a facial form captivatingly poised between the attractions of the racial Other and of the European self. The film-maker and photographer Leni Riefenstahl was able to accommodate the Maasai of Kenya within the same aesthetic that in previous times, in Nazi Germany, she had employed to glorify the Aryan.

Petrus Camper's quest for universal standards of beauty is enjoying a revival, thanks to computer graphics and the new wave of biological determinism. The economy of images has been globalised along with everything else; the 'United Colors of Benetton' advertising campaign is only the most self-conscious example of the pervasive assertion that all races can discover beauty in all other races. Indeed, the acme of beauty may be attained by the combination of 'all the ideals of all the different continents', in the words of the fashion designer Vivienne Westwood, who analyses the features of the celebrated model Naomi Campbell typologically: 'an icon of a face, African oval with slanty eyes and flat planes; she's part Chinese; her nose looks a bit Arab to me'.[2] This last, incidentally, seems to echo the old typological belief that East Africans were a 'Nilo-Hamitic' mixture of African and Arab.

By and large, though, people still believe that beauty is in the eye of the beholder, and that the beholder's gaze is refracted through an ethnic prism. The underlying, largely unvoiced assumption is that, particularly when it comes to actual relationships with real people, one will principally be attracted to those of one's own race.

This is a special case of the general principle that it is natural to prefer one's own kind; special because it doesn't arouse suspicions of prejudice. If you hear somebody on a television or radio programme invoke the principle, you may be confident they are about to call for immigration controls, cuts in the foreign aid budget, or some other political expression of ethnocentrism. Yet nobody would argue that it is necessary to find people of other races physically attractive in order to be anti-racist. In that sense, beauty provides a space for guilt-free prejudice.

Beauty, then, is a domain in which normal moral and social considerations regarding race are suspended, or at least modified. To a certain extent this is also true of athletic ability, in certain fields of which 'elongated' Africans are also acknowledged to be outstanding. Yet popular explanations of the international success of Ethiopian and Kenyan runners betray unease about the permissibility of talking about physical differences and ethnic groups in the same breath. To some, it is obviously the result of a lean build and long legs. To others, it shows that people from poor countries can outcompete those from wealthy ones when money, in the form of expensive training programmes, is not a prerequisite.

When the Ethiopian runner Abebe Bikila won the 1960 Olympic marathon, barefoot and barely trained, he created a powerful image for modern sports culture. The idea of the barefoot athlete, running across the landscape without artificial assistance, has both classical associations, evoking the ideal of the pure contest of bodies in the original Olympic games, and natural ones, evoking the primordial mobility that took early humans out of Africa and around the world. The socioeconomic explanation also asserts an essential human equality underneath global inequalities of wealth. As world champions, people from one region of Africa stand for the entire continent, making a symbolic demonstration of Africa's tragically unrealised potential.

A redemptive device of a quite different sort is used by the evolutionary biologist Jonathan Kingdon in his book *Self-Made Man And His Undoing*.[3] 'If the blistering rays of a tropical sun bouncing off waters and beaches of the Banda Sea once weeded out countless numbers of children and adults, there is a certain justice that their most distant descendants should vindicate the cost of that selection with medals,' he observes. We can only speculate about whether the ancient humans on the fatal shore would have felt it was all worth it if they had known there was Olympic gold at the end of the line. But there is absolutely no justice in evolution, any more than there is purpose – which, in effect, is what Kingdon is implying, albeit only as a rhetorical conceit.

Kingdon's account of human evolution emphasises the diversity to be found both within the species as a whole, and within groups defined by the physical markers of race. While rejecting both racism and the traditional race concept, he insists that populations may enjoy 'genetic advantages'. These permit him to reconcile a modern populationist evolutionary perspective with remarks on racial character that hark back to an older idiom. 'Few would argue with the statement that black men and women are outstanding runners, strong and resilient athletes with exceptional speed and stamina,' he observes, and the praise of muscle fibre seems to speak also of moral fibre.

The molecular anthropologist Vincent Sarich also asserts that there are truths about race and sport which are self-evident. 'If you can believe that individuals of recent African ancestry are not genetically advantaged over those of European and Asian ancestry in certain athletic endeavours, then you probably could be led to believe just about anything,' he told students taking his introductory anthropology course at Berkeley.[4] 'On the average a black athlete can outjump a white athlete,' he noted. 'That fact says nothing about two individuals, one "white" and one "black".' However, if the averages are different, then the white group will contain a disproportionate number of poor jumpers, and excellence in jumping will be the almost exclusive preserve of the black group. 'There is no "white" Michael Jordan (one of the best basketball players to ever play the game) nor has there ever been one.' Where it counts, white men can't jump.

These observations, made in the first lecture of a controversial course, established the principle of meaningful racial differences – even with the races in inverted commas – and embedded it within the principle of statistical distribution popularised a few years later by Richard Herrnstein and Charles Murray as *The Bell Curve*. Here, the importance of sport as a precedent for race differences in other domains was explicit and provocative. The example was given as an illustration of the course's central message, that our natures are the product of evolutionary history rather than of contemporary environments.

Sarich raised it as a counter to the tendency in American universities to proscribe claims of innate differences between groups. Specifically, it was a reaction to a controversy at the University of Maine, reported in the *New York Times*. The university's president had said that research showed blacks were better suited genetically than whites for certain sports. He received a collective admonition, from students, faculty, state officials, and the chancellor, who considered that the observation was 'at the very least insensitive and showing poor judgment'. The president of the Afro-American Student Association was reported to have objected both to the implied stereotype, and to what he saw as the belittling of his achievement as a football player.

Amby Burfoot, an editor at the US magazine *Runner's World*, has identified an article which appeared in *Sports Illustrated* in 1971 as a turning point in the discussion of the issue. The criticism it received set the tone for the next twenty years. Writing in 1992, Burfoot noted that physical differences were not even mentioned in a series of articles about black American athletes which *Sports Illustrated* had run the previous year. He added that while researching his own piece, he had tried to consult a leading American sports scientist. 'Go ahead and hang yourself,' the scientist had told him, 'but you're not going to hang me with you.'[5]

Burfoot, a former winner of the Boston Marathon, argued that the issue had been muddled by a concentration on 'country club' sports such as tennis, where the relative absence of blacks could readily be traced to social inequality – or plain discrimination. Citing the example of Bikila, who had not needed expensive coaching, or even shoes, he suggested that no scientist interested in performance differences could invent a better sport than running.

Like Sarich's observations on basketball, this claim needs to be qualified. Jumping is surely as 'natural' an activity as running. Yet blacks do not dominate international high jump competitions to the extent that they dominate basketball. African–Americans excel in sprint events; Africans do not. No sport can be assumed to be the sort of culture-free system a scientist would ideally like.

Nevertheless, Burfoot is probably right to suggest that no sport is better suited to scientific investigation than running. He describes unpublished work by the Swedish exercise physiologist Bengt Saltin, who visited the highland area that has produced many of Kenya's best runners to compare the performances of Swedish distance runners with those of local people. Despite their national tradition of success in endurance events, the Swedes were beaten by hundreds of Kenyan youths.

Saltin found physiological differences in the muscle tissue of the two groups. The Kenyans had more blood capillaries surrounding their muscle fibres, which contained more of the energy-generating units known as mitochondria, and were smaller, perhaps improving the diffusion of oxygen from the capillaries into the mitochondria. After exercise, their muscles contained less ammonia than those of the Swedes, suggesting that their metabolism was more efficient in using fat as fuel, rather than glycogen and protein, which would improve their capacity for endurance.

Exercise, of course, is based on the principle that environmental intervention can have a dramatic effect on the body. Saltin considers that the Kenyans' physiological advantages may result from a lifetime of running and walking at altitude. But Peruvians and Tibetans do that too, responds Burfoot, and they don't feature strongly in the distance-running records. Why the Kenyans?

He answers that they must have the perfect combination of genetic and environmental factors, noting cultural traits of stoicism, aggression and male dominance. By contrast, Owen Anderson, editor and publisher of *Running Research News*, sees no need to invoke genetic factors. A physiologist by training, he attributes the Kenyans' achievements firstly to the superiority in fitness they enjoy over Westerners. 'They have to run to live,' he remarks.[6] Secondly, their economic opportunities are limited; whereas, thirdly, Kenya provides a superb support system of training and employment for its runners. In his account of a visit to Kenya, Anderson stresses above all the rigours of the athletes' training schedules, which he records in minute and painful detail.[7]

As for the Peruvians and Tibetans, Anderson argues that being born at an altitude of 8,000 feet or more will stimulate the development of anaerobic rather than aerobic exercise capacity, rendering the individual unsuited to distance running. The highland Kenyans, by contrast, are born at altitudes of 6–7,000 feet, which is ideal for the development of aerobic capacities.

None the less, there is an intuitive appeal to the idea that the 'elongated Africans' of Kenya are physically suited to distance running. Might it not be that when individuals of a type adapted to hot and arid conditions occupy a highland region, the interaction of genes and environment happens to produce ideal runners' bodies? Anderson agrees that the East African atheletes look like perfect runners from head to foot, all aspects of their physical form seeming as if designed for the purpose. But he argues that we do not actually know just what physical attributes are ideal for running. Even the role of stature is not obvious. Although all the Kenyan runners are slender, one or two of them are only around five foot tall. Perhaps when Westerners look at East African athletes, the ideal types they see are largely projections of Western aesthetic ideals, like the fashion-plate Maasai.

Whatever the truth about African athletic capacities, however, Amby Burfoot highlights the fundamental point that Africans are not all alike. His main contribution to the debate may not have been grasping the nettle in the first place, but popularising the idea that a West African and an East African may be as different from each other as each is from a European. Of all the continents, Africa encompasses the greatest variability in its human populations. To take visible characteristics alone, the tallest and the shortest people on Earth are found there; as are people with the thickest and the thinnest lips; while African cranial dimensions span 80 per cent, and nasal width 92 per cent, of the entire human range.[8]

East Africans' sporting achievements therefore say nothing about the performance of West Africans; but the point was not grasped by two academics who contributed to a review of the race question of the *Canadian Journal of Sport Sciences* in 1988.[9] They rightly identified the lack of African success in sprinting as a problem for explanations based on physical racial differences, but then wrecked their case by comparing the absence of 'American Blacks' from the distance events in which 'African Blacks' excel. Bizarrely, their definition of blackness extended to Moroccans and Tunisians.

Burfoot notes that Americans are far more inhibited about discussing the race and sport question than Europeans. 'Why Can Blacks Run Faster Than Whites?' asked a headline in the *Sun*, the leading British tabloid daily paper, after the black British athlete Linford Christie won the 100 metres world championship at Stuttgart, in a race against seven other athletes, all of them also black, 'It's all in genes say experts' was the answer given underneath.[10]

The expert cited was Gordon Bosworth, a sports physiotherapist, who was quoted to the effect that blacks have heavier bones which are less vulnerable to injury; they have less fat under the skin because less heat insulation is needed in hot countries, and therefore can build up more 'fast twitch' muscle fibre. But the rest of the article was devoted to emphasising that physical advantages mean nothing without the individual dedication and effort needed to become a world champion – and to escape a deprived upbringing in a racist society. The moral of the story was the same as that of the parable of the talents in St Matthew's Gospel: what matters is not what you are given, but what you do with it.[11]

The *Sun* laid rather more weight on science than Vincent Sarich did in his remarks on sport and genes. He suggested deleting 'research' from the Maine university president's statement, and substituting 'common sense'. Claude Bouchard, who edited the *Canadian Journal of Sport Sciences* collection of papers, was considerably more circumspect than Sarich.[12] Bouchard, of the Physical Activity Sciences Laboratory at Laval University in Quebec, concluded his preface with the observation that physical differences 'are probably quite limited in comparison to the individual differences seen within each race'. Rather than illustrating the effects of genetically determined bell curves, as Sarich would maintain, the patterns of racial success at the top end of sporting competition 'may have more to do with opportunities and socio-economic conditions'. From Sarich's evolutionary-populist viewpoint that presumably means Bouchard will believe anything – even that a white man can jump.

The occupational statistics and the anthropological findings converge on that question. Although African-Americans comprise about eleven per cent of the US population, they fill more than seventy per cent of the country's professional basketball rosters.[13] Robert M. Malina's review, part of the collection edited by Bouchard, demonstrated that six out of ten studies showed black males doing better than white ones in the long jump, and seven out of eight studies showed blacks doing better in the vertical jump.[14] On the other hand, other studies have found that whites do better in free-throw shooting.

The film *White Men Can't Jump*, short on science but long on formulas, is none the less generally consistent with the literature, notably the introductory scene in which Woody Harrelson's character wins respect from the blacks by beating Wesley Snipes's character in a free-throw contest. A sociological explanation is at hand to explain this particular performance difference, though. It has been suggested that white kids hone their basketball skills in the comfort of their own suburban driveways, shooting at the net without any pressure from competitors. Blacks, on the other hand, have to compete for space in overcrowded urban courts, where they need to be able to keep the ball from being taken by attacking players, and to shoot under pressure. In the end, Woody Harrelson proves himself able to mix it up with his street-tough opponents. Much of the movie is about the ability of whites to cross racial barriers by cultural adaptation, and at its climax, Harrelson manages to falsify the title of the film. The message is that of all modern sport: moral determination, not biological determinism.

The obverse of the claim that white men can't jump is the idea that black people can't swim. Champion swimmers are typically of European descent, while swimmers of African descent are conspicuous by their absence. A common suggestion is that this may be because blacks have denser bones, which tends to be regarded as a throwback to the old, discredited race science. But the finding is well established, as one of the papers in the Canadian collection notes. A higher concentration of body calcium and bone mineral has been found to be accompanied by larger muscle mass, which may have stimulated the bone growth indicated by the mineral concentration.[15]

To suggest that denser bones are part of the reason why blacks do not excel at swimming is not to say that they are bad swimmers. As they grow up and explore their capabilities, young athletes will tend to concentrate on the areas in which they do best. Thus a minor difference in average ability may result in a much more visible difference at elite levels of performance. Nevertheless, any claim of black physical inferiority naturally touches a raw nerve.

The very different sporting arena of British football helps illustrate why this is so. One component of the package of prejudices which hindered the careers of black footballers in the 1960s and 1970s was the claim that black players could not cope with the cold British climate. Research has indeed shown that men of African descent have greater difficulty than men of European descent in adjusting to low temperatures. When immersed in iced water, the blacks' fingers have a far lower average temperature than that of the whites' fingers.[16] But the claim about black footballers was more than simply an assertion about physiology. It conflated physical traits with temperamental qualities, implying that blacks lacked the moral fibre to survive in the tough British game. As a statement about character, it supported the more general charge that black players lacked 'bottle', meaning courage and determination. It also hinted that blacks and the chilly British way of life were basically incompatible. Typically, black players would be deployed as wingers, but they were not admitted to the midfield 'engine room' of the team. Midfield players need to be tough, to win the ball, and they need to be able to orchestrate the play. In other words, they need courage and intelligence, precisely the qualities that black players were considered to lack. Out on the wing, they were in a position where they could do some good but little damage; they were not trusted in positions of responsibility. They were acknowledged to have 'flair', a quality with strong connotations of instinct, whereas white players doing similar things might be credited for their intelligence. Looking abroad, the British were prepared to make an exception for the legendary Pelé and his colleagues of African descent, but their genius was understood as a specifically Brazilian thing, not a black thing.

The advance of Afro-Caribbean British footballers has eroded these prejudices. An individual case in point is that of John Barnes, who performed disappointingly as a winger in the English national side, but went on to play a major tactical role in midfield at Liverpool FC.[17] (There is a striking parallel to this issue in American football, where the quarterback is a sort of one-man midfield, and is almost always white.)

Nevertheless, the stereotype still seems to be embedded in certain areas of football culture, notably management. It is another example of the durability of the Romantic concept of race, though in this case it is expressed in folk rather than scientific discourse. The Romantic vision of race united external appearance, bodily characteristics, temperament and mental capability. When racial prejudice links cold adaptation to moral fibre and intelligence, it is hardly surprising that black people are often suspicious of claims about physical difference, even when these appear to stand in isolation from any other ideas about black characteristics.

There are other reasons for caution about the evidence of physical race differences. It is not always easy to work out the connection between an observed physical difference and an observed difference in performance; for example, why African–American physiques should give them an advantage in sprinting. Muscle power, necessary for sprinting and jumping, increases with cross-sectional area, yet African–Americans tend to have relatively slender calves.[18]

Nor is it easy to fit specific performance differences into a global pattern. Reviewing studies from around the world, extending to South Africans and to Andeans, to Malaysians and to Lapps, Marcel R. Boulay, Pierre F. M. Ama and Claude Bouchard fail to find any reliable evidence in favour of innate differences. There are certainly variations in performance, but these can be explained by environmental factors such as the degree of familiarity with the test apparatus. Overall, the average work capacities of different ethnic groups seem remarkably similar.[19]

These researchers and their colleagues in the Physical Activity Sciences Laboratory have also discovered striking differences between the muscle tissue of blacks and whites, the former having a higher percentage of type IIa fibres. Enzyme activity in blacks was also markedly higher – from 40 to 76 per cent; this, like the muscle fibre pattern, appeared extremely likely to be a genetic effect. On paper, these differences ought to give blacks an advantage in short-burst sports events. But their work output was the same as, or worse than, that of the whites.

In sum, then, it seems reasonable to suppose that human groups do vary in some physiological traits that are implicated in athletic ability, and that some of this variation is genetic in origin. But that does not warrant jumping to conclusions about the relationship between physiology and performance, since sport is a cultural phenomenon, not a biological one, and since the physiological evidence itself is more complicated than it at first appears.

To make it possible to talk about physical differences without reiterating the old racist subtext, physiology needs to be uncoupled from psychology. Black people need to feel that discussion of African physical traits does not reinforce the stereotype that associates blacks with the body and whites with the brain. Sportspeople may enjoy great wealth, social status, fame and personal satisfaction, but they do not control countries, financial systems, or the development of technology. If praise of black athletic achievement inhibits the development of black intellectual potential, the leading lights of black communities may be enormously popular, but they will lack real power.

One consequence of admitting that human groups may have significant bodily differences, however, is that it implies the possibility of differences in the brain. Besides all the other reasons for unease, people are wary of talking about physical differences because they have an intuitive sense that doing so puts them at the top of a slippery slope. If it could be plotted on a graph, it would look like a bell curve.

5

Fifteen Points

WHAT BIOLOGY HAS done with race bears a certain resemblance to what physics has done with particles. It has extracted the contradictions and ambiguities hidden within the solid, discrete certainties of traditional race theory, and founded the populational concept upon them. Examined at the equivalent of the quantum level, races are revealed as fluid, elusive and paradoxical, if not illusory.

Meanwhile, psychology is content to remain in the clockwork universe. The questions it is set to address are to a large extent social ones, and therefore they are framed in social terms that may well dissolve at the biological level. The requirement for statistics on populations to be ethnically divided is as strong as it ever was – if not more so, now that the demand is made most insistently by radical anti-racists. Psychologists tend to take socially constructed ethnic categories as a given. Some then try to extract biological truths from these imperfect schemes.

The overwhelming mass of these endeavours revolves around the idea of intelligence. David Wechsler, a prominent psychometrician, defined it as 'the capacity of an individual to understand the world about him and his resourcefulness to cope with its challenges'. The psychometric project is based on the belief that this diffuse and polymorphous quality can be captured by intelligence tests in the form of an intelligence quotient, or IQ. This is argued to be the measure of a factor called general intelligence, or g. Sometimes known as Spearman's g, after Charles Spearman, who first defined it, it is as fundamental to those who believe in it as the other g, that stands for gravity, is to physics.

In other, 'hard' disciplines, claims like these may seem absurdly pretentious. Discussing the IQ concept, the psychologist Charles Locurto refers to the Nobel physics prizewinner Richard Feynman's dismissal of much of modern psychology as 'cargo cult science'.[1] During the Second World War, the inhabitants of a number of Pacific islands had grown accustomed to the flow of manufactured goods from the industrial world that had been borne down from the skies by aircraft supporting the war effort. Seeking to maintain the supplies after the end of hostilities, the islanders built runways, lit beacons, and posted men wearing wooden headphones with bamboo antennae. Similarly, Feynman observed, psychologists 'follow all the apparent precepts and forms of scientific investigation, but they're missing something essential because the planes don't land.'

If there is truth in this, then the therapeutic variety of psychologist would immediately diagnose the subdiscipline of intelligence studies as a massive case of overcompensation. The most striking characteristic of the IQ debate, as it has unfolded since Arthur Jensen first set out his hereditarian stall in 1969, is the sheer volume of text generated: 123 pages for Jensen's inaugural paper in the *Harvard Educational Review*, 'How Much Can We Boost IQ and Scholastic Achievement?', 786 pages for his 1980 volume *Bias In Mental Testing*, and so on through to the 845 pages of *The Bell Curve* in 1994. This massive edifice of theory is part of an older tradition, dating back to Francis Galton, who first proposed the idea of general intelligence. Galton made major advances in the science of statistics because of his need to structure the anthropometric data which he devoted enormous energy to gathering (not to mention his sideline interests in quantifying beauty and the power of prayer). In the first decades of the twentieth century, the pioneers of intelligence-testing made further contributions to statistical method in their quest to turn their data into a theory.

Their methods had a decidedly un-Galtonian origin. Mental testing arose from the decision that French children of especially low ability would be taught in special schools, instead of being excluded from the education system altogether. In 1905, Alfred Binet introduced a package of tests intended to identify such children. His approach was to use a variety of tests in order to gain a measure of underlying mental capacities which could be expressed as a single figure. From 1908, tests were graded according to the lowest age at which Binet considered a normal child could cope with them; the age rating of the most difficult test a child could manage became his or her 'mental age'. The decision on whether a child should be sent to a special school was based on the difference between the mental age and the real age. But the younger the child, the more serious a given difference is, so the relationship is better expressed by dividing mental age by actual age. The German psychologist who pointed this out, William Stern, multiplied the result by 100 to give a whole number, and the intelligence quotient in its familiar form.

Alfred Binet not only regarded the original application of intelligence tests as the extent of their utility, but described the belief that IQ scores are destiny as 'brutal pessimism'.[2] Such pessimism flourished on the other side of the Atlantic, where IQ was adopted as a means of grading all levels of intelligence. Henry Goddard, who adopted Binet's methods with enthusiasm, coined the term 'moron' to describe the targets of the Binet tests, those children who were possibly but not obviously unsuited to normal schools. He believed that they should be kept in institutions, where they would be prevented from reproducing and passing on the gene he supposed was responsible for their condition. Outside institutional confines, they formed the backbone of the delinquent classes; because of their lack of intelligence, they lacked the moral sense to avoid lives of crime or prostitution. Goddard was perturbed to find apparently large numbers of morons among the shiploads of poor immigrants from Europe – mostly Jews, other East Europeans, and Southern Europeans – arriving at Ellis Island.

The flow of immigrants from those regions was curbed by the Immigration Restriction Act, passed in 1924 after Congressional debates which frequently turned on IQ data. Eugenics, Nordicism and intelligence-testing combined to project the idea that the health of the American nation depended upon the maintenance of Anglo-Saxon supremacy, which meant that steps had to be taken to maximise the proportion of the American populace that derived from Northern European stock.

The lines of this sketch are derived from Stephen Jay Gould's trenchant historical analysis, *The Mismeasure of Man*, which appeared in 1981. Between the publication of Jensen's original paper and that of *The Bell Curve*, probably no single text did more to shape public perceptions of the intelligence question, and it has therefore become a *bête noire* for hereditarians. (Before going any further, it is probably necessary to point out that the labels 'hereditarian' and 'environmentalist' denote orientations rather than absolute positions. Nobody of any significance in either camp believes that intelligence is determined entirely by genes or entirely by environment.) Gould's message is 'that determinist arguments for ranking people according to a single scale of intelligence, no matter how numerically sophisticated, have recorded little more than social prejudice'.[3]

Arthur Jensen dismisses Gould's account as the biased selection of a few ' "bad apples" '.[4] Mark Snyderman agrees, contending that 'the majority of early testers were engaged in a legitimate scientific enterprise'. He describes the connection posited between contemporary IQ hereditarianism and the eugenicists and racial supremacists of the early decades of the century as 'guilt by association'. Nowadays, he claims, the major flaws evident in the early years of testing have been eliminated.[5] This is the Vienna *Rassensaal* defence: today's science is methodologically sound and ideologically untainted, while yesterday's was basically clean as well.

Using their improved techniques, the modern IQ hereditarians have spun a vast web of theory, within which two principal axes can be identified. One seeks correlations between IQ and other phenomena. The other seeks evidence supporting the contention that intelligence, as measured by *g*, is largely inherited. As social scientists, they are overwhelmed by any rating that passes the half-way mark. The concept around which their discipline revolves, correlation, is another of Galton's innovations. It represents the degree of association between two sets of figures – the tendency for high (or low) values of one variable to be associated with high (or low) values of another – and is measured on a scale of 0 to 1 if positive, or 0 to −1 if negative. In the social sciences, as Herrnstein and Murray note, correlations higher than 0.5 are rare.[6]

Against this yardstick, the correlations between IQ and school grades look unarguable, at 0.65 for elementary schools and 0.55 for high schools. Not much less impressive is the correlation of 0.45 between the IQs of husbands and wives, said to be the strongest association within any marital parameter tested. Physical co-ordination and cognitive ability show a correlation of 0.35, according to a US Department of Labor study. Weaker, but still positive, is the 0.20 correlation between the measured IQ of 84 female students and figures estimated by assessors on the sole basis of photographs of the women's faces. Other positive associations have been reported between IQ and height, speed of speech, altruism, and sense of humour.[7]

Hereditarians like Daniel Seligman, a former editor of *Fortune* magazine, who presents these associations in his popular book *A Question of Intelligence*, are inclined to see them as evidence of the wonderful power of *g* to index mental phenomena, or at least as tantalising glimpses of a mental landscape through which IQ tests are the royal road. A more sceptical eye might see at least some of these as a reminder of the fundamental caveat of statistics: correlation is not the same as causation. In some cases, such as that of height, a causal relationship with intelligence does not seem intuitively likely. In others, such as school performance, it seems sensible; and it is also intuitively apparent which one causes the other. Once such an example is accepted, other causal relationships come to seem more plausible. Perhaps intelligence enables people to see beyond their own immediate interests and to appreciate the benefits of helping others. Perhaps intelligent people get jokes that stupid ones don't understand. Perhaps a woman's face is the window of her mind.

On the other hand, a correlation may be as near to perfect as is possible in the real world, while the two variables it connects are perfectly unrelated. Gould gives the example of the perfect correlation between his age and the expansion of the universe.[8] In other cases, the variables may share a common cause. Wealth, for example, might cause children to grow taller than their poorer peers, and to have higher IQs as a result of better education. The meaning ascribed to correlations depends on the interpreter's overall view of the phenomena in question. A hereditarian may consider wealth unlikely to have much effect on intelligence; an environmentalist may find it appealing as a causal factor. There will probably be positive correlations between their opinions on this question and their respective political orientations.

Similarly, hereditarians may feel that investigations into the relationship between IQ and the sense of humour may yield results of value. On the other hand, faced with Gould's charge that IQ serves the interests of the elite classes by justifying the inferior status of the poor and ethnic minorities, they are likely to agree with Daniel Seligman that this is a 'wild proposition'. Gould's opponents frequently describe him as a 'Marxist' – although his view of the role of chance in history is a bourgeois deviation from dialectical materialism if ever there was one.

For intelligence theory, the most important correlations of all are the internal ones between scores on the various different tests. Probably the two most highly regarded test batteries are those developed by David Wechsler, the Wechsler Intelligence Scale for Children and the Wechsler Adult Intelligence Scale (WAIS). Within the revised version of the latter, the WAIS-R, the correlation between the Information subtest and the Vocabulary subtest is 0.80, while that between Information and Object Assembly is 0.45.[9] Charles Spearman observed that although the correlations were very variable, they were all positive. He developed a technique, known as factor analysis, which generated a general factor, g, from the whole set of test scores. This, he claimed, was the measure of general intelligence.

The nub of the anti-IQ argument is that g is a statistical artifact and nothing more. It is easy to generate positive correlations from nonsensical data, as Peter Schönemann shows, while incidentally demonstrating that mathematics can be sarcastic. His imaginary example is that of a Great Society, in which Greatness is worshipped in all its forms – 'large families, high income, big houses, long names, tall stature, large shoe sizes'. Each of the two tribes who live in the Great Society practise Pseudometrics, 'the science of measuring undefined variables'. Using tests based on the number of great aunts, the number of letters in the month of birth, and shoe size, the pseudometricians of each tribe come to predictably opposite conclusions about which tribe is Greater.[10]

Arthur Jensen rejects this satirical allegory as 'sophistry'. Positive correlations between tests exist because there really is such a thing as general intelligence underlying them, he insists. The differences revealed by the tests correspond to the genuine differences in ability that are observed in real life; the appropriate retort to Schönemann is that of Samuel Johnson, who responded to doubts cast on the existence of external reality by kicking a stone and exclaiming, 'I refute it thus!'

Even if there is a meaningful relationship between ordinary experience and test results, however, that does not prove the existence of *g* as Jensen and his followers know it. In *The Mismeasure of Man*, Stephen Jay Gould demonstrates that all roads of factor analysis do not necessarily lead to *g*. ('In my entire career, I have never had such difficulty rendering a technical subject in layman's terms,' he later observed.)[11] If *g* is just one of a number of possible ways of representing the data, he argues, then it cannot have the importance, or the external reality, that the Jensen school attributes to it. This line of criticism in fact dates back to L. L. Thurstone, who in the 1930s rejected *g* in favour of a group of independent 'primary mental abilities'.

Today, the most prominent theorist of the tradition is Howard Gardner, who identifies seven intelligences: linguistic, musical, logical-mathematical, spatial, bodily-kinesthetic, intrapersonal and interpersonal.[12] At the most obvious level, seven dimensions are better than one because they provide more information. A child's school report lists grades for each subject, rather than a single figure representing the average of the individual scores. A single figure would probably give a reliable indication of the child's general level of performance, since bright children tend to do well in most subjects, and dull ones tend to do badly across the board. In the jargon, there is a positive correlation between performances in different subjects. But a single figure would conceal important information about the child's strengths and weaknesses.

In the same way, multiple intelligences serve to highlight the differences between types of skill. They acknowledge that understanding how to get on with people is significantly different from knowing how to solve an equation. (At the same time, our mental flexibility allows different individuals to do things in different ways. Autistic people, for example, may use their reasoning ability to gain the insights into other people's emotions that they cannot grasp intuitively.)

Multiple intelligences also seem far superior to g in their power to express the range of abilities implied by Wechsler when he defined intelligence in terms of understanding the world and coping with its challenges. Intelligence tests have been developed to meet the needs of modern technological society, but intelligence itself was not. IQ tests provide very useful information about performance in the cultures which developed them, but it does not follow that they can tell us much about how people in other cultures – including our distant ancestors – have understood the world and coped with its challenges. If you start from IQ, you are obliged to take Western Man as the model for humanity as a whole, throughout its history. If you begin by trying to consider the full range of human experience, across cultures and evolutionary time, you may be more inclined towards a multiple model of intelligences. Such models also seem to be in tune with current thought among the scholars who have recently begun to call themselves evolutionary psychologists. They tend to see mind as the product of 'modules', packages dedicated to specific faculties, rather than of a single general-purpose processing unit.[13] As with g, however, there is a risk of reification: of taking a statistical artifact to correspond to a real object.

A psychology of multiple intelligences might well be a more humanistic science than the cold doctrine of g. But it would not necessarily eliminate the race question. If the entire g school was obliged to convert to a multiple intelligence model overnight, reports of racial differences in one factor or another would doubtless begin to emerge within a week. Before much longer, hereditarians would be happily reinventing the old race gallery of racial types with distinctive patterns of both physical and mental traits, all backed by batteries of statistics. And even if comparisons between ethnic groups were made without any reference to genes or environment, race relations would probably not benefit greatly from claims that one group is more musical or more skilled in interspersonal relations than another.

For the moment, however, *g* remains the pivot of racial psychology, and the principal role of multiple models is that of a weapon for opponents of hereditarianism. When Gould reiterated the factor-analysis critique in his *New Yorker* review of *The Bell Curve*, however, Charles Murray retorted that the 'factor-analysis argument was so irrelevant to the state of knowledge even when it was published that it provoked only perfunctory comment in the technical literature'. He concluded: 'we have learned not only that *g* survives statistical scrutiny and is related to many social and economic outcomes but also that it is associated with brain function at the neurological level. Not bad for a statistical artifact.'[14] Not conclusive, though. Multiple intelligence models might produce factors more strongly associated with neurology or the fortunes of individuals. And such associations do not prove that *g* is more than an artifact.

For its adherents, *g* is not only ubiquitous, but is one of the principal determinants of the structure of modern society. Occupations, and occupational status, correspond to particular IQ bands. A rough but clear hierarchy emerges from studies by Linda Gottfredson and Robert Gordon, leaders of the Project for the Study of Intelligence and Society. Truck drivers are found in the 86–112 range; firefighters, police officers and electricians between 91 and 117; teachers and estate agents from 108 to 134; doctors and engineers upwards of 114.[15] Thus a child's future station in life can be predicted better by IQ tests than by anything else, and the essence of social superiority is intellectual superiority.

It is also identified as the key to social legitimacy. Illegal occupations can be fitted into this scheme: the IQ of the typical criminal is around 90, according to James Q. Wilson and Richard Herrnstein.[16] Robert Gordon considers that the high crime rate recorded among black Americans can be ascribed to their low intelligence. Reviewing a book by another scholar, the Edinburgh psychologist Christopher Brand sanguinely observed that admitting evidence for the causal role of low IQ allowed the difference in black and white crime rates to be handled 'unprovocatively'.[17]

In *g*, psychologists have found a universal factor to replace typological race. Instead of a society in which garage mechanics are overwhelmingly recruited from the Mediterranean race, IQ takes the

Western world into a modernised, numerised mode, where mechanics are overwhelmingly derived from a particular stratum of intelligence. Hierarchy is no longer asserted by wordy ruminations on national character, but read off from a numerical scale.

Like many trends flourishing in the world today, IQ theory has a somewhat archaic flavour. It still has roots in the great heyday of technocracy, when organisations were rationalised by treating the humans within them as quantifiable components like any other. Latterly, the Enlightenment of rational planning has been occluded by a Romantic reaction, in the form of an individualistic, therapeutic, emotional ideology of 'human resources'. This, in addition to the 'liberal bias' of the media incessantly decried by the hereditarians, is perhaps a significant factor discouraging the public from embracing g wholeheartedly.

Popular discourse in recent years has also come to emphasise talk and feelings, rather than performance. 'Bonding' of various types is now understood to be fundamental to social and personal health. We see ourselves as primates, grooming, courting and cheating. Theories from the scientific domain enter into dialogue with this perception: they propose that the evolutionary benefits of intelligence lay in the improved ability to interact with other individuals in a primate group. A cynical moral can be drawn from this account – that our intelligence is 'Machiavellian', being configured in a way that helps us plot and deceive – and so can a feminist moral: that chatter and gossip are not the froth of human communication, but the driving force of intelligence. In both these respects, the theory speaks to the mood of the time. While the natural evolutionary myth behind IQ is based on performance – the hunter deploying his cognitive abilities in a complex physical environment – the notion of socially driven intelligence emphasises feelings and interpersonal relationships. It directs us towards primate society, rather than machine society.

Whether society is felt to be shaped more by its original primate nature, or by the emergent machine properties it has developed, there is a universal conviction that ethnic divisions are fundamental to its structure. If society is also committed to meritocracy, then it is acutely important that individual achievement is independent of ethnic divisions. Otherwise ethnic groups will become characterised by high or low achievement, and ethnic diversity will turn back into ethnic hierarchy.

The message from the IQ data, mostly collected in the United States and interpreted within a North American framework, is unequivocal. When the two great terms of the racial equation, black and white, are compared, the black average is about fifteen points below the white one – or sixteen, or eighteen, according to the hardliners. Like the old eugenicist and racial supremacist traditions from which it dissociates itself, Jensenism is preoccupied not with the middle but with the extremes.

As Murray and Herrnstein have once again explained, most of the scores in a set of data will be close to the mean, or average, value of the whole set of data. The further from the mean, the fewer scores there will be. If ten coins are tossed, for example, the mean number that come up heads will be five. This will be the commonest score; the least frequent scores will be zero or ten. If all the scores are plotted on a graph, the resulting curve is bell-shaped. This is known as the 'normal distribution'. The extent to which scores are scattered around the mean value is expressed in a measure called the standard deviation. This is the square root of the variance, which is defined as the average of the squares of the amounts by which individual scores deviate from the mean. In an ideal normal distribution, 68 per cent of cases fall within one standard deviation either side of the mean; 95 per cent within two standard deviations either side; and 99.7 per cent within three standard deviations.

Fifteen points is about one standard deviation. With mean IQ defined as 100, the mean black IQ is around 85, on the lower borderline of Linda Gottfredson's truckdriving class. Only about fifteen per cent of blacks will be over the centreline, and only about two and a half per cent will reach the threshold for physicians and engineers. In a population of 30 million, however, 100,000 still get into the top 125+ bracket, Herrnstein and Murray point out, as if this ought to cheer up African-Americans.[18] The same tactic, it may be recalled, was used in the 1951 Unesco Statement.

On the other side of the mean from this six-figure elite are the six million blacks implied by the rule of the curve to have IQs below 75, widely considered to mark the borderline of mental retardation. This amounts to one in five blacks, while the equivalent figure for whites is one in twenty.[19] Daniel Seligman spells these implications out, whereas Murray and Herrnstein, more finely attuned to their public's sensibilities, have a more judicious sense of what to leave unsaid. Nevertheless, there is no doubt about the black-white IQ gap. The controversy is over what it really measures, and what causes it.

In the days before cultural politics, those in search of an environmental explanation for the poorer performance of subordinate groups often felt they had to look no further than economics. Arthur Jensen argued that this is not enough in the case of the black IQ deficit. Socioeconomic status – a broader measure than income alone – typically accounts for about a third of the gap, which not only persists, but actually widens as socioeconomic status rises: the differences between the scores of the wealthiest blacks and whites are larger than those between the scores of the poorest. Jensen and his school consider that they have also disposed of other possible environmental explanations, such as the difference between Black English Vernacular and the standard English of the tests and the testers. And if this is not enough, they then reverse the arrow of causation, making explicit the suggestion that IQ determines status, rather than the other way round.

Jensen notes that blacks do better on some tests than others. But these, he says, are the ones that are less 'g-loaded' – in other words, those that require less intelligence. This is most neatly demonstrated by a subtest known as digit span, which involves no more than repeating back a series of numbers. The twist lies in the racial difference in performance on the forward version – in which the subject repeats the numbers in the order they were presented – and the backward version, which requires the subject to reverse the order. The gap between blacks and whites is found to be twice as large on the backward test as on the forward. The g-loading – the amount of cognitive processing required – is higher on the backward test. For Jensen, this is part of the body of evidence which proves that the difference between black and white IQs is a difference in *g*. And as time has passed, he says, he has become more pessimistic that anything can be done about it.[20]

The Jensen school has been challenged on its own ground over its rejection of environmental explanations. But even if the Jensenite case is allowed to stand, its model leaves one environmental possibility open. Jensen has dubbed this 'Factor X', which may represent the effects of racism, working in a way which cannot be pinned down by controlling conventional environmental variables. For Jensenites, Factor X is primarily a rhetorical device. It is admitted to be a theoretical possibility as a way of turning the formalities of debate into a means of ridiculing the opposition. Profoundly committed to their methodology, and conservative in their political outlook, hereditarians regard the idea of an unmeasurable factor of racism as absurd. In fact, they rate X about as highly as environmentalists rate *g*.

The X factor, however, is a more recognisable force nowadays than it was when environmental factors were conceived in numbers, usually dollars. It can be seen as the sum total of everything – slavery, segregation, the visible blip in alert levels as whites pass black males on the sidewalk, the 'nihilism' identified by Cornel West, or the fact that you can be a middle-aged professor in a suit, like Cornel West, but a midtown New York cab still won't stop for you if you are black.[21] This is subjective alienation, and also more than that: it is the cumulative effect of a history which is currently in a phase of crisis.

Jensen recognises a number of its manifestations, such as slavery, but proceeds to demand that they have 'some psychological generality (i.e. are not confined to minority groups of sub-Saharan African descent)'. The whole point, though, is that the experience of these groups in America is historically specific, deriving from an experience of slavery and subsequent oppression not shared by other groups. It would not be possible to separate out the components of this Factor X, since their interactions must be infinitely more complex than the simple additive model by which variables are usually controlled. Therefore they cannot meet Jensen's requirement of empirical testability, and are excluded from his model, but that does not mean they do not exist. Conversely, the fact that *g* thrives within his theoretical structure does not prove that it exists outside.

For hereditarians, a powerful counterargument to X arises from tests not of intelligence, but of 'performance' – specifically, of reaction time. The underlying supposition, originally mooted by Galton, is that intelligence is based on the speed at which the nervous system operates. The task is to press a button when a light comes on. Performance is divided into reaction time, the interval between when the light comes on and when the subject's finger starts to move, and movement time, the time it takes for the finger to reach the button. This most simple of processes is thus separated into mental and muscular components, and the former is found to correlate more strongly with IQ scores than the latter. Whites react more quickly than blacks; blacks move quicker than whites. Thus within a fraction of a second, the stereotypical identification of whites with the brain and blacks with the body is upheld.

There is no question of background knowledge to muddy the result, since the subjects do not even have to be able to count. And the possibility that blacks are less tuned in to the tests seems to be ruled out. As Herrnstein and Murray observe, how could a subject be unmotivated during the reaction phase, but apparently motivated during the movement phase? Reaction time studies give the impression of operating at a level below the cloudbase of cultural factors. No matter how indirect their evidence is, the message they send is that the difference between blacks and whites is one of neural efficiency, not attitude.

As set out in *The Bell Curve*, the hereditarian interpretation of the reaction times seems decisive. There is no suggestion that environmentalists have any answer – apart, of course, from the basic objection that *g* is meaningless. But they do, and their reservations are sketched out in none other than the *Mankind Quarterly*, as part of a critical commentary on Richard Lynn's theories.[22] Ian J. Deary points out that while there is indeed a correlation between reaction times and IQ scores, it is typically a modest 0.2 or so. This is too low, he argues, to have any implications for the debate over genetic and environmental influences on intelligence. And while processes underlying reaction time may underlie IQ scores too, the converse is also argued; that high IQ improves reaction time.

Reaction time and similar measures remain a secondary, though significant, strand in the hereditarian argument on IQ and race. Its central theme is the claim that the heritability of IQ is too high to allow an environmental explanation of the black-white disparity. Heritability is the proportion of the variation of a trait that can be ascribed to genetic variation. (It is therefore measured across populations, and cannot separate the respective contributions of genes and environment to the makeup of an individual.) At the minimum, hereditarians insist on heritability levels for IQ of 0.4 or 0.5, which still leaves the door open for the environmental explanation. But they generally place more weight on much higher figures, of around 0.7 or 0.8.

These numbers are derived from studies of twins, who come in two varieties. Dizygotic, or fraternal, twins develop from separate eggs and sperms. Monozygotic, or identical, twins grow from a single fertilised egg that divides into two. Their genotypes, or genetic codes, are therefore identical. Nature having controlled this variable, they are thus potentially hot experimental material. But to realise their potential fully, they must be raised apart.

This happens infrequently, and only a handful of studies have been conducted upon them. Of these, the work of Sir Cyril Burt has been discarded, as his figures are considered too neat to be true. This was recognised before he was posthumously accused of outright fraud and the invention of collaborators, and will probably continue to exclude his data from the literature despite his successful rehabilitation in the early 1990s. All the earlier work has in any case been largely superseded by the Minnesota Study of Twins Reared Apart, conducted by Thomas Bouchard, who has cornered the market in monozygotes. Averaging several results, Bouchard arrives at a heritability figure for IQ of about 0.70.[23]

Bouchard's finding enjoys an unusual degree of public influence because his project is one of those rare scientific endeavours that is capable of seizing the popular imagination. As well as IQ, he and his colleagues have also run the gamut of twin personality, assessing religious tendencies, political conservatism and liberalism, job satisfaction, leisure interests and marital stability. They conclude that in these respects, their fifty pairs of separated monozygotes are about as similar as they would have been if raised together. As if this blanket dismissal of environmental influences in general, and the family in particular, were not enough, the data are accompanied by extraordinary anecdotal reports. One pair of separated twins found, for example, that they shared not only a genotype but a habit of flushing the lavatory before as well as after using it, and a fondness for sneezing loudly in lifts. Stories like these are more commonly found accompanying claims of paranormal phenomena. These are enthusiastically dissected by the 'Skeptic' scientific rationalist movement, but it has yet to conduct what would amount to an internal inquiry.

Given a favourable cultural climate, the Bouchard findings and their accompanying tales – simple, memorable, televisable, and bizarre – are exactly the right kind of account to establish the validity of genetic determinism in the public mind. The inclusion of IQ in the package is thus all the more significant. Coming where it did in the history of twin and intelligence studies, the Project was also in a position to strengthen its findings by addressing criticisms of earlier work. Among these was the objection that there are often considerable similarities between the homes in which separated twins are raised, not least because they are often brought up by relatives or friends. Sometimes they are in contact with each other as they grow up.

Bouchard claims that such similarities only added 0.03 to the IQ correlations, and denies that contact between twins has any significant effect on it at all. Nevertheless, it seems a good idea to retain a degree of caution about the twins who take part in studies. Twin culture is one of conspicuous similarity, as displayed at twin conventions, and thanks to Bouchard, Minnesota has become an international centre of twinhood. Like any other stories, twin accounts are selected and shaped by their narrators, including the twins themselves.

Whatever heritabilities are calculated, however, there remains a fundamental objection to comparisons between groups. The heritability of a trait within a group does not explain the differences between groups. With weary exasperation, Stephen Jay Gould repeated the argument in 1994 that he had made in the early 1980s: the heritability of height might be extremely high in both a poor Third World village and a wealthy First World suburb – taller parents would produce taller children in both places. The richer population might be taller than the poor, but this does not preclude the possibility that a better-nourished generation in the South might in the future grow as tall as its distant peers in the presently developed world. The correlation between the heights of parents and children might remain just as strong, but the heights themselves might change.

Not in America, the hereditarians retort. Since the 1960s, a number of projects have attempted to boost the IQ of deprived children. The improvements have generally been transient or equivocal, reassuring conservatives that social problems cannot be ameliorated by public spending. But there are signs that the gap between black and white is narrowing. Herrnstein and Murray note that, since 1969, improvements in black scores have narrowed the gap recorded by the National Assessment of Educational Progress from 0.92 of a standard deviation to 0.64. This may reflect rising living standards among African-Americans in general, while excluding the effects of impoverishment and social crisis in the black underclass, who will probably not get as far as the tests. Consistent with this trend are several 1980s studies which found IQ disparities of seven, ten or twelve points, rather than fifteen. The psychologist Robert Plomin has also suggested that the heritability of IQ is in decline, on the basis of a level of 0.50 he calculated in 1980.

At the same time, IQ scores themselves may be on the way up. Unsettlingly for psychologists whose worldview is based on the conviction that g is an entity with its feet firmly planted on the theoretical floor, IQ test results appear to be anything but earth-bound. There is nothing surprising about a 'secular trend', as it is called, in IQ. Like height and weight, IQ is expected to rise from generation to generation, and the tests are 're-normed' at intervals to adjust for this. The effect was recognised long before James Flynn, of the University of Otago in New Zealand, began to survey it around the world. (Herrnstein and Murray have now catchily dubbed secular trend in IQ the 'Flynn effect'.)

Flynn's challenge to mainstream IQ theory – and to those who are pessimistic about the possibilities for boosting IQ and scholastic achievement – lies in the magnitude of the effect observed. Between 1950 and 1980, he has found, scores rose by eleven points on verbal tests, fifteen points on the Wechsler test battery, and eighteen points on a test package called Raven's Progressive Matrices. These figures happen to match those found for the gap between blacks and whites – suggesting, if taken at face value, that African-Americans in 1980 had the same intelligence level as whites thirty years before. The effect was also evident in twenty other countries, the most dramatic example coming from the Netherlands and France, where one generation appears to be 20 to 25 points ahead of its predecessor. Between 1972 and 1982, Dutch 18-year-olds gained 8.67 points. Eventually, Flynn came to the conclusion that IQ tests 'do not measure intelligence but rather a correlate with a weak causal link to intelligence.'[24]

Flynn is perhaps the only antagonist whose arguments are conceded by hereditarians to pack a punch. Jensen singles him out as a 'respectable' critic.[25] For once, IQ theorists have to clutch at environmental factors – Richard Lynn proposes nutrition – to explain the effect away. Not even the most convinced hereditarian could imagine that the genes of young Dutch or French people could transmute themselves that fast. Flynn himself concludes simply that IQ tests 'cannot bridge the cultural distance that separates one generation from another'. And if the tests cannot bridge the gap between generations, they may be incapable of bridging other cultural gaps, such as those between ethnic groups.

As all concerned agree, the only way to settle the question of race and IQ would be to conduct a controlled experiment in which newborn babies of one race were handed over to parents of another and followed through childhood. Actually, this would still not include antenatal or perinatal environmental influences, but the point is a rhetorical one – except in the case of Juan Comas, who proposed such a study in the early 1960s during the controversy over the *Mankind Quarterly*.[26]

Transracial adoptions in the real world – where the adopting parents in the sample are all white – have been studied by the behaviour geneticist Sandra Scarr and her colleague Richard Weinberg. At the age of about seven, the black adoptive children attained scores as good as those of white adoptees, averaging 106.

Ten years later, adopted children whose biological parents were both black scored little better than black children raised in black families.

For hereditarians like Herrnstein and Murray, the follow-up study affirmed that race will out, obliterating any small influence environmental factors may temporarily enjoy. Scarr herself – who has dismissed critics of hereditarianism like Stephen Jay Gould and Leon Kamin as 'eccentrics' – none the less still argues that environment has a significant role. All the lines of her extensive research, she said in 1987, have given evidence against the racial genetic hypothesis and for the importance of individual variability.[27] Even her more recent research, she and her colleagues maintained, produced 'little or no conclusive evidence for genetic influences underlying a racial difference.'[28] The roots of the inequality, they suggest, may lie in the period before the child goes to live with its adoptive parents. The sorry condition of the adolescent may also result in part from the everyday experience of being black in a white-dominated society, which gets tougher as the child approaches adulthood, and loses whatever protection from the world that childhood affords. Scarr also criticises Herrnstein and Murray for failing to address evidence from a study she and Weinberg conducted among African-American adolescents, comparing their IQ scores to blood proteins which indicated how much of their ancestry was African. There was no relationship between the two.[29] The IQ performance of African-Americans is therefore related to their social categorisation as 'black' in America, not their biological links with Africa.

The climate for the IQ of black people often seems more congenial outside the United States. In Bermuda, Scarr and Weinberg found that black children have scores equal to the white norm in the United States at the age of two, and just one point lower at four; by the time they reach grade 3, they are two years above US whites in vocabulary, reading and mathematics.[30] In Britain, a study of cognitive and educational attainment in 2,000 children, aged between seven and fifteen, found that those of Afro-Caribbean descent often performed as well as those of indigenous or other European stock. More significantly for the race theory of intelligence, the Afro-Caribbean children scored higher than their peers of

Indian subcontinental descent, confounding the racial expectation that Caucasoids will outperform Negroids.[31]

Besides transracial adoption, there is one other social phenomenon that, as Flynn points out, provides direct evidence on what effect growing up in a white environment has on the IQ of black children. Like Unesco Man, the research material for what later became an environmentalist trump card was the direct, if informal, result of the aftermath of the Second World War. In 1961, a researcher named K. Eyferth published the results of a survey of illegitimate German children fathered by American servicemen from the occupation forces. The average IQ of those with black fathers were virtually identical to that of those whose fathers were white.[32]

Herrnstein and Murray note this study as 'intriguing' if 'inconclusive'.[33] The reaction of another commentator shows that it is not just liberals who may find race research hard to take. 'The most serious methodological defect in these studies is biased sampling. The black and white American fathers of illegitimate German children are not likely to be representative of the black and white American soldiers stationed in Germany; nor can the German girls mating with black soldiers be assumed to be equivalent to those mating with white soldiers,' expostulated Robert Nichols.[34] A generation before, a girl could have got herself sterilised or institutionalised for mating with a black man, which was just the kind of behaviour Henry Goddard's generation of psychologists took as evidence of 'feeble-mindedness'.

Even without remarks like Nichols's, racial psychology is deeply vulnerable to the charge that, in Gould's words, it records little more than social prejudice. That it has recorded prejudice in vast amounts is not in doubt; the question is whether it has done much else. For the answer to be yes, IQ scores must measure intelligence, they must be relatively stable, and they must be comparable across different cultures. If there is a real faculty in the human mind corresponding to g, then IQ tests are measuring intelligence. As Gould argues, since other statistical techniques account for the data just as well, but do not require g, there is no reason to assume that g is anything more than a statistical abstraction. As Flynn argues, soaraway IQ scores are far too volatile to provide a firm basis for a genetic theory of racial intelligence differences, and demonstrate that relatively minor cultural differences prevent comparisons between the scores of different groups. Unless the hereditarians can dispose of all these question marks – and build a body of direct evidence, such as that from adoption studies, which gives them unambiguous support – it seems wise to regard their theoretical edifice as a castle in the air.

The hereditarians, who can boast a formidable array of scholarship, naturally think otherwise. After a quarter of a century, neither camp can claim to hold sway throughout society as a whole. To assess how the debate on race and intelligence has affected society, which is a quite different issue from that of which side is right, it is necessary to consider where the competing schools' zones of influence lie. The first question is, who actually believes in IQ?

One of the most persistent complaints from the Jensenite camp is that of media bias. To read the liberal press, they say, you would think that no reputable scholar places any weight in IQ tests, and that none would entertain the idea that there are innate differences in intelligence between ethnic groups. The *New York Review of Books*, a resolute bastion of opposition to IQ theorising, is the journal they most love to hate. Their complaint finds some support from Adrian Wooldridge, author of a book on intelligence testing in Britain. He observes that the *New York Review* 'gave the impression that the only respectable debate on the subject was between different versions of the Marxist orthodoxy, as Stephen Jay Gould played Mutt to Richard Lewontin's Jeff'.[35]

Environmentalist conviction is not the preserve of the left, however. In 1991, editorialising on 'America's wasted blacks', *The Economist* acknowledged that 'deep in their minds some whites have begun to think what their ancestors thought, that blacks are genetically inferior in the traits that count for economic success, and that this is proven by the fact that blacks have lost ground as discrimination has retreated.' Rejecting this conclusion, it affirmed that 'the underclass is disproportionately black by historical accident, not genetics: blacks migrated to the northern cities at an unlucky moment, just as manufacturing jobs were disappearing to take new form in the suburbs. They are trapped there by crime, drugs, unemployment and poor education.'[36]

There is more behind this analysis than simple generosity of spirit. In the 1980s and 1990s, *The Economist* has epitomised (and proselytised for) the worldview of neoliberal economics. This depends on a fundamental optimism about human potential, because it relies upon the ability of individuals to take care of themselves, thereby relieving the state of burdensome social responsibilities. Its economic models consider individuals as entities that are basically undifferentiated and therefore flexible: dynamic economies need to be able to recruit and dispose of labour as necessary, drawing on a pool of human resources regardless of sex or race.

As a universalistic ideology that considers its terms of reference sufficient to explain most things of any importance about society, it does not set much store by factors distant from economics, such as those of biology or culture. The alienated cultural perspective of African-Americans is dismissed as what Marxists used to call false consciousness. In *The Economist*'s opinion, it is not white oppression that is holding African-Americans back, but their own conception of themselves as victims, and their belief that the state should be there to help them.

For neoliberal optimism to be sustained, this has to be true. Although a genetic glass ceiling for black people would not invalidate the neoliberal position, since the rule of the bell curve permits a few individuals to cross the threshold into the elite, it would be hard to sustain the spirit of optimism with such a substantial group basically ruled unfit to compete. What is true for committed neoliberal economists is also generally true for American society, which has invested so heavily in the idea of self-improvement, as well as for other Western societies.

Amid this fervent and pervasive belief in human potential, psychology is a hotbed of pessimism. What the lay media say about intelligence is not a fair reflection of expert opinion, and the hereditarians have figures to prove it. In the mid-1980s, Mark Snyderman and a political scientist named Stanley Rothman sent out questionnaires to 1,020 psychologists of various stripes – educational, metrical, developmental, counselling, cognitive and industrial, along with the behaviour geneticists – with a common interest in intelligence.[37] Of the 661 who replied, 58 per cent expressed a belief in some sort of general intelligence, while 13 per cent expressed a preference for the idea of multiple intelligences. 'On the whole,' the authors commented, 'respondents seemed to believe that intelligence tests are doing a good job measuring intelligence, as they would define it.'

Of those who considered themselves qualified to answer, only half felt able to estimate the heritability of intelligence among white Americans; they produced an average figure of 60 per cent. When asked about the heritability of the gap between black and white IQ scores, only 15 per cent of respondents considered that it was entirely due to environmental factors. Fourteen per cent did not respond; 24 per cent thought that insufficient data were available to form an opinion. An ultra-hardline fringe of 1 per cent agreed that genes explained the whole of the gap. The biggest group, 45 per cent, believed that the difference is a product of both genetic and environmental variation.

Although the results showed a wide range of views and a great deal of circumspection about IQ, the bottom line was that more than half endorsed g, and nearly half of those expressing an opinion agreed that the IQ gap between blacks and whites is partly due to genes. Accompanied by an analysis of press articles about intelligence, the survey helped turn a Pavlovian whine about media bias into an element of a conscious bid for hegemony. The book has been widely cited in Jensenite literature, but was ignored by reviewers, which the authors' views on the media presumably led them to expect.[38]

If Snyderman and Rothman were the scouts, Herrnstein and Murray were the assault force. The book that eventually appeared as *The Bell Curve* in 1994 was trailed long before its completion. It was eagerly awaited by a small coterie of scholars who had developed their racial preoccupations in the privacy of the academic margins, and were now hoping the time was right for the marginal to become mainstream. It was not clear in advance whether Murray and Herrnstein would tackle race directly, or simply furnish some conservative journal with the pretext for raising the issue.[39] These academics wanted race back on to the agenda. They wanted it to be legitimate once more.

The only new thing about *The Bell Curve* was the *Zeitgeist* of the moment at which it appeared. It was a consummately judged attempt to mould itself to prevailing opinion, in order to effect a major change in public discourse. Flattering and cosseting its lay reader with unremitting artfulness, it is an impressively lucid – and large – work of popularisation. For bluffers, there are chapter summaries in italics which, as the authors helpfully point out, cut the book down to about thirty pages. For those who are gripped by feelings of intellectual inadequacy at the sight of a graph, the text is held with a sure hand at a level which supports without seeming to condescend. One effect of this is that there are no obvious bumps or holes that would prompt the reader to step back and ask awkward questions. It all looks seamless.

Herrnstein and Murray also highlight evidence that contradicts their thesis, thereby creating a reassuring sense of authority and objectivity. They have to do this if they are to be credible, since their thesis includes propositions that have been marginalised, and thus have the aura of the crackpot fringe about them. But, as noted in the discussion of reaction time above, there is more to the opposition than they admit. Stephen Jay Gould accuses them of structural sleight of hand, in which they bury the evidence that refutes their case in the appendixes.

Another critic, Charles Lane, discovered something else untoward in the book's bulky nether regions. As well as five direct citations of articles from the *Mankind Quarterly*, Lane found citations for seventeen researchers who had contributed to the journal, of whom ten had served in an editorial capacity, either as editors or members of the Advisory Board.[40]

Lane's article in the *New York Review of Books* also detailed connections between hereditarian science, race science, and a body called the Pioneer Fund. Established in 1937 by a textile machinery magnate named Wickliffe Draper, the Fund was intended to 'improve the character of the American people' by encouraging higher reproductive rates among the descendants of 'white persons who settled in the original thirteen colonies prior to the adoption of the constitution and/or from related stocks'. Its other goal was to foster research into, and disseminate information upon, heredity, eugenics, and 'race betterment'.[41] Draper's principal collaborators were the scientists Harry H. Laughlin and Frederick Osborn. Laughlin drew up the Model Eugenic Sterilization Law on which the Nazi Law on Preventing Hereditarily Ill Progeny was based. Osborn called the German action 'the most exciting experiment that had ever been tried'.

Today, the Pioneer Fund has become the financial base of research into human inequality. When public agencies refuse to provide funds, Pioneer is there to sustain hereditarian and race research. 'There are no major projects in America into inherited intelligence other than the ones we fund,' Pioneer's president, Harry F. Weyher, claimed in 1992. That changed, however, when Robert Plomin and Gerald E. McClearn were allocated federal backing to search for genes that influence mental functions.[42]

The Pioneer Fund disburses around a million dollars a year.[43] In the second half of the 1980s, its main beneficiary was Thomas Bouchard's twin study project, which was given more than $500,000.[44] Figures presented by Adam Miller in *Rolling Stone* indicate that between 1971 and 1992, Arthur Jensen received more than one million dollars. Roger Pearson got $787,400 during this period; close behind him was J. Philippe Rushton, of the University of Western Ontario, who has received $770,738. Richard Lynn benefited to the tune of $325,000, while Linda Gottfredson is listed as receiving $267,000 and Robert Gordon $214,000.[45]

Among the other beneficiaries was the late William Shockley, who won a Nobel prize for developing the transistor and subsequently devoted himself to the proposition that races were colour-coded for intelligence, like electrical resistors. More recently, Pioneer funds have been allocated to Michael Levin, a philosophy professor at the City College of New York, whose public speciality is standing up in front of African-American audiences and telling them that they are criminally inclined because of their lack of intelligence.[46]

The Pioneer Fund's particulars have been noted before, both in academic publications and the mainstream media, without visible effect on its *modus operandi*.[47] In the past, its standard response was to deny accusations of racism, and quietly slip back into the shadows. The *Bell Curve* furore subjected Pioneer and its kindred spirits to more intense scrutiny. At the same time, though, it seemed to give them a new confidence about appearing in public. Harry Weyher came out fighting, with a letter to the *Wall Street Journal* that, as well as rejecting 'the various charges of nefarious conduct and intent', made a strong claim for positive legitimacy. The researchers funded by Pioneer, he asserted, 'are very much in the scientific mainstream'.[48]

As well as a list of distinctions boasted by some of these scholars – Top Five placings on the world citation chart for living psychologists, Guggenheim awards, fellowships of the American Psychological Association, and so on – Weyher referred to a statement which appeared in the *Wall Street Journal* on 15 December 1994. Entitled 'Mainstream Science on Intelligence', it was Jensen, Snyderman and Rothman in the form of a manifesto. Drafted by Linda Gottfredson, its theory was Jensenist; its claim of legitimacy rested on the assertion that the views expressed were mainstream, as quantified by Snyderman and Rothman.[49] Presenting itself as a corrective to misleading commentary on *The Bell Curve*, it was an explicit bid for hegemony in the public sphere of discourse.

The statement was signed by 52 professors, 'all experts in intelligence and allied fields', fifteen of whom, Weyher noted, had received Pioneer Fund money. Fifty were based in the United States, the other two, Hans Eysenck and Richard Lynn, in the United Kingdom. The roll-call included Arthur Jensen, Thomas Bouchard, Robert Gordon, Sandra Scarr and Robert Plomin. There was Raymond Cattell, who was born in 1905 and has been warning of the threat facing the nations of the West from genetic deterioration for most of the century. In 1987, he proclaimed a scientistic religion based on group selection and eugenics; for the future, he advocated the deliberate splitting of humanity into different species, through genetic engineering or the colonisation of the solar system.[50] Like Lynn, who sympathises with his views, he features on the *Mankind Quarterly* masthead. There was Vincent Sarich, indicating that a molecular anthropologist can be considered as occupying an 'allied field' if he speaks up for hereditarianism. And there was J. Philippe Rushton.

This added up to a remarkable coalition. At one end there was Sandra Scarr, the behaviour geneticist who doubts that genes matter in racial IQ differences. At the other were Richard Lynn and J.

Philippe Rushton, who believe that Mongoloids and Caucasoids are the products of evolutionary selection for intelligence and social discipline. The *Mankind Quarterly* and the most prominent of the race science hardliners were included in the 'mainstream'.

The statement concluded with the observation that the research findings 'neither dictate nor preclude any particular social policy'. For some of the signatories, however, the data point in a very clear direction. Seymour Itzkoff, a professor of education, outlined his views in a book that appeared around the same time as *The Bell Curve*. In *The Decline of Intelligence in America: A Strategy for National Renewal*, he called for immigration curbs and the ending of welfare programmes. Robert Gordon has called for a campaign to discourage the unintelligent from breeding. Richard Lynn supports abortion on demand 'as a way of reducing the fertility of less competent and less intelligent people'.[51]

Francis Galton, in whose footsteps these scholars are proud to tread, defined eugenics as 'the science which deals with all the influences that improve the inborn qualities of a race; also with those that develop them to the utmost advantage.' In recent years, the concept of eugenics has made a comeback, but has been redefined as a matter of individual choice. Warnings are issued that people will increasingly be faced with eugenic decisions as reproductive and genetic technologies advance. The notion of a collective enterprise, for the betterment of the race or the nation, is not expected to come into their calculations, however. This is the result both of the extreme individualism of Western society in the late twentieth century, and of the exclusion of biological race from public discourse. By and large, the possibility of a new eugenics in the proper sense just did not occur to liberal commentators. (Singapore's eugenic policies were regarded as just one of the city-state's many idiosyncrasies; China's leanings in the same direction have received little more than passing attention in the West.)[52] Yet eugenics is undoubtedly back, thanks to the hereditarians and to *The Bell Curve*, which leads its readers through a vision of future 'dysgenesis', and then tiptoes off to an idealised smalltown America which can be regained, it promises, if welfare is abolished.

Herrnstein and Murray claim that people's private thoughts on race are very different from what they say in public. This may well be true, though it is difficult to gauge what the public has been thinking about race differences in recent years, because it is not often asked. In 1990, a University of Chicago National Opinion Research Center survey found that 53 per cent of non-blacks believed black Americans to be less intelligent than whites, but did not ask whether they thought the difference was innate.[53] When the 1986 Race and Politics Survey presented respondents with the statement that 'blacks are born with less ability', only 6 per cent agreed. The researchers, Paul M. Sniderman and Thomas Piazza, contrast this with the 23 per cent of whites who agreed to the proposition when surveyed in the San Francisco area less than ten years before.[54] A British survey conducted in 1991 found less than 20 per cent of whites prepared to agree that white people are more intelligent than blacks. Less than 50 per cent disagreed, however, leaving a group of 'don't knows' bigger than that for any of the accompanying questions about racial attitudes.[55] This suggests that white opinion remains open to persuasion on the matter.

Whatever the state of public opinion, Herrnstein and Murray assert that 'the fascination with race, IQ and genes is misbegotten'. They mean, first, that what counts is an individual's capabilities, and an individual of any ethnic group may have a high IQ. Many critics did not feel this was an entirely adequate solution to the racial intelligence question. 'What's the difference between thinking that the black male next to me is dumb and thinking there's a 25 per cent chance he's dumb?' asked Alan Wolfe, one of twenty commentators given space in the *New Republic* after Andrew Sullivan's decision to publish a piece by Herrnstein and Murray caused almost his entire editorial staff to revolt.[56]

Second, they mean that the environment can be as difficult to counteract as the genes. Sometimes, genetic shortcomings are very easy to overcome: with spectacles, for example, as Gould points out.[57] But it is central to the *Bell Curve* project to show that interventions have only a limited effect in boosting IQ. Projects such as Head Start can certainly seem as ineffectual as attempting to empty a lake with a bucket. There are two possible conclusions. One is that the poor need a 'Marshall Plan', or indeed a social transformation that would join them up with productive employment and guarantee healthcare through socialised provision. Another is that, since such an option runs counter to prevailing orthodoxy, nothing can be done.

The Bell Curve's interpretation of the environmental dimension also helps make it politically palatable. 'Just as by denying the role of race in *Losing Ground* Murray made himself seem a reasonable, race-neutral scholar, so by denying the importance of heredity in *The Bell Curve* he seems a nice, non-racialist fellow,' observed Mickey Kaus.[58] Personal criticisms of this kind were directed almost exclusively at Murray, Herrnstein having died around the time that the book was published.

Just how much Charles Murray cares about race is open to question. *Losing Ground*, the book which made him famous in 1984, was about the growth of the 'underclass', and what he has latterly called the 'new rabble' seems to remain the principal object of his concern. In his demonology, the welfare mother, of whatever colour, looms much larger than the homeboy. Alan Ryan, keeping up the *New York Review of Books'* fire against the forces of hereditarianism, depicted *The Bell Curve* as the result of Herrnstein's obsession with race and Murray's obsession with the welfare system.[59]

That, however, may be precisely Murray's significance: as the public voice which draws together indisputably mainstream political concerns, and those of scientists who uphold the old belief that race is everything.

6

The Higher Latitudes

THE QUESTION OF racial intelligence, conceived for many years in terms of black and white, began to acquire a third dimension in the 1970s. Japanese economic success induced soul-searching in the West, and Richard Lynn began to gather figures to support his favoured explanation. While a professor of social psychology at the Economic and Social Research Institute in Dublin, he became interested in the social psychology of economic growth. He started to suspect that the differences in growth rates between nations were the result of differences in national intelligence.

In 1977, he published the first of many papers asserting the superior intelligence of the Japanese and other East Asians. Even with a modest-looking mean of 103 – a figure he now gives as a reasonable estimate – the geometry of the bell curve indicates a significant advantage for the Japanese in the elite regions where Lynn and his colleagues believe national success is determined.[1] Lynn's IQ data combined with the scholastic results of Asian Americans, particularly among the new generation of settlers, to send the message that people from the far side of the Pacific Rim were doing better because they were cleverer.

Was this the result of their ancient culture, or something more ancient still? In one of Garry Trudeau's 'Doonesbury' cartoon strips, a European American teenager demands to know the secret of Asian American scholastic success. 'I just study,' replies her schoolmate Kim.

'No way,' she retorts, 'I tried that once.' Environmental influences thus excluded, 'some genetic edge' has to be behind it. 'You guys are some sort of superrace, aren't you?' 'We mean you no harm,' Kim reassures her. 'We seek only computers for our young!'

According to figures gathered by a group including Thomas Bouchard, and cited by Lynn, the heritability of work motivation is around 0.4. He adds, however, that the growth of welfare payments may have sapped the motivation of the black underclass in the United States.[2] But from the beginning, he suspected that genes were responsible for the East Asian advantage, and he cannot accept that social factors are a credible alternative explanation.

A principal reason is that he believes one immigrant group to be in basically the same position as any other. Most are the victims of prejudice, but some do well and some do badly. He dismisses the idea that 'because blacks' great great grandfathers were slaves this has in some way caused them to have lower IQ's today', as if the Civil Rights Act had been passed in 1864, the year after Lincoln proclaimed emancipation, rather than in 1964.[3]

Lynn believes that the success or failure of migrant groups, and ethnic groups in general, is determined by their own innate capacities. Thus Jews and East Asians do well everywhere despite oppression, which is therefore not the reason that blacks do badly everywhere. Voluntary migration may be seen as a sign of intelligence, since pioneer settlers – but not subsequent arrivals – must be an especially resourceful breed. Lynn believes that this effect can be detected among the Irish, lending weight to the premise of Irish stupidity on which so much of English humour depends. Surveying the data, he finds an IQ average of 95: he suggests that emigration, such as that which took place after the potato famine, skimmed off the intellectual cream of the Irish people, leaving a dim and dysgenic mass behind. His views were given prominence in the London *Sunday Telegraph*, after a court awarded a machinist from Northern Ireland damages for having to listen to Irish jokes made by his workmates.[4]

Lynn also favours another hypothesis about dysgenesis which, if true, would certainly have impinged upon the Irish. The argument, one of Francis Galton's, is that in Catholic countries, the brightest boys were steered into holy orders and prevented by the rule of celibacy from passing on their genes. This has been proposed as the explanation for the low intelligence of Hispanics in America – though Lynn thinks the main factor is probably hybridisation with Native Americans.[5]

This seems at first glance to be an argument that would appeal to Nordicists. But conversely, Jews were said to have benefited from thousands of years of turning their brightest boys into rabbis, and encouraging them to have large families. The argument seems to enjoy widespread if underground currency: it surfaced in the wake of a provocative article in the London *Spectator* about the alleged domination of Hollywood by a Jewish clique.[6] Another eugenic proposition endorsed by Lynn is that 'the Jews have been subjected to intermittent persecutions which the more intelligent may have been able to foresee and escape'.[7] So every pogrom has a silver lining.

Lynn's genetic hypothesis of racial intelligence is based on the triple division of humankind into Negroids, Caucasoids and Mongoloids. For it to stand up, the rank order with Negroids at the bottom and Mongoloids at the top needs to be consistent around the world. At the upper end, he is challenged by James Flynn, who claims that the difference in means between East Asian Americans and European Americans is illusory (though Flynn accepts that the former have lower scores on verbal skills). Lynn argues that the surveys available are flawed, as is Flynn's model of intelligence.[8]

In Africa, by contrast, Lynn is much readier to give the benefit of the doubt. His theory predicts that sub-Saharan Africans, as pure Negroids, will score lower than African-Americans, who are Negroid-Caucasoid hybrids. Sure enough, he extracts a mean figure of 70 from the IQ data available. This produces a simple and striking hierarchy, with Mongoloids somewhat more intelligent than Caucasoids, Caucasoids much more intelligent than Negroid-Caucasoid hybrids, and pure Negroids trailing far down the field.[9]

The African scores, gladdening as they may be to white supremacists, actually push the genetic hypothesis towards an absurd extreme. Given that the threshold of mental retardation is widely set at 70, the implication is that half the black African population is mentally retarded by Western standards, and millions in the lower reaches of the range are below the level at which they can perform

the basic operations of life unassisted. Europeans may have traditionally regarded Africans as childlike, but the idea of the white man's
burden surely did not extend to feeding and dressing a significant
proportion of his colonial charges.

The African figures appear to be a demonstration of James Flynn's
argument that IQ tests are limited in their ability to bridge cultural
distance. Flynn himself cites a study by Jensen, in which black
schoolchildren in a Georgia town were found to have an average IQ
of 71. Their scores declined as they got older, however, and on this
basis Flynn estimates that the maximum average IQ of the adult black
population was 63.5. Yet the blacks of this unnamed town appeared
quite normal, and in no need of benevolent white guardianship.[10]

Five of the eleven African studies Lynn cites were conducted in
South Africa, between 1929 and 1991. (The 'pure Negroid' mean in
the 1929 survey was only 65.) All but the first were conducted under
the period of National Party rule that began in 1948. The attitude of
the apartheid regime to black education is indicated by the fact that,
during the mid-1970s, the state spent more than fifteen times as much
on each white pupil as on each black pupil.[11] The results of
underfunding included shortages of books, shortages of classrooms,
sometimes solved by putting two classes and two teachers in one
room, and shortages of teachers, sometimes tackled by getting one
teacher to look after two classes.

Conditions had improved somewhat by 1989, when the study
Lynn regards as most reliable produced a mean figure of 69. Yet its
author interpreted the results in quite a different way from Lynn.
Interviewed by Charles Lane for his *New York Review of Books* article,
'The Tainted Sources of "The Bell Curve" ', Ken Owen ascribed
poor black African performance to bad schooling and difficulties
with English. The scores, he said, 'certainly cannot' be used as an
indicator of black African intelligence in general. Lane also quotes
from a 1988 review by South African psychologists, 'Test Performance of Blacks in Southern Africa', which concluded that 'it would
be rash to suppose that psychometric tests constitute valid measures
of intelligence among non-westerners', and questions 'whether
there is any point in even considering genetic factors as an additional
source of variance'.[12]

The kernel of Lynn's hypothesis is located elsewhere, however, in his Oriental studies. It is based on the inadequacy of g, though intelligence theorists would be unlikely to put it that way. To get the best of both worlds, reconciling Spearman's g with Thurstone's model of multiple intelligences, Sir Cyril Burt and Philip E. Vernon developed a hierarchical model in which g contains two secondary qualities, verbal and visuospatial intelligence. The latter is the basis of mathematical ability, and it is this which Lynn regards as the secret of East Asian success. When the general intelligence score is unpacked, it turns out to have concealed a verbal component of about 96, which rises to about 100 as children move through their school careers, and a visuospatial one of around 106, which remains constant. The Oriental pattern of intelligence, visuospatially strong but verbally weak, is why Asian Americans become engineers and scientists, but not lawyers.

Lynn considers that this characteristic pattern of intelligence renders environmental explanations highly unlikely. Environmentalists, he argues, have to suggest factors which would boost one sort of ability but leave another slightly depressed. They also have to account for a pattern of childhood development in which the East Asian advantage is slow to emerge. They have to propose a reason for the way in which the high visuospatial score appears around the age of six, but stays constant through to sixteen, instead of rising together with the verbal score. And they have to explain why the intelligence pattern persists in Asian Americans several generations after migration from East Asia.[13]

Ironically, the Japanese themselves maintain a strong traditional belief in *hakushi*, or the *tabula rasa*, regarding the child's early years as critical to its development.[14] Such store is set by child-rearing practices that they are maintained among third-generation Japanese Americans who have become Americanised in most other ways; and aspects of these practices seem conducive to the 'Asian intelligence' pattern. According to Joy Hendry, author of *Becoming Japanese: the World of the Pre-school Child*, the principal goal of child-rearing is the avoidance of anxiety. Japanese mothers tend to soothe their children, rocking them gently, where American mothers tend to stimulate physical and verbal responses by patting and talking to theirs. Hendry compares the outgoing American personality with the Japanese ideal of silence and stoicism.

In the Japanese child, the dominant effect of this style of mothering is said to be *amae*, the tendency to depend on love and indulgence. Perhaps this, rather than a slower process of physical maturation, might explain the delay before Japanese children's intellectual capacities are revealed within the formidable Japanese school system. And an important factor in the success of East Asian schools, according to Harold W. Stevenson, is a deep-rooted Chinese and Japanese belief in the human potential for achievement, given effort. Americans, on the other hand, are much readier to accept what they suppose to be inherent limitations, like the Euro-American teenager in the Doonesbury cartoon.[15]

Cultural factors can thus be invoked to explain all the characteristics identified by Lynn as rendering environmental explanations unlikely, except for the key element of strong visuospatial skills. Lynn argues that if schooling were responsible, then one would expect to see cumulative rises in both verbal and visuospatial skills. An alternative scenario might be that the Japanese mode of child-rearing delays the expression of both verbal and visuospatial abilities, but while verbal skills take a long time to develop (perhaps slowed by the use of ideographic characters in writing), visuospatial ones are precipitated by basic skills acquired at the beginning of children's school careers. The persistence of a bias towards visuospatial abilities might simply be the result of their early establishment.

The point is not that any of these proposed effects are more or less likely than genes to be the cause of the observed intelligence pattern; nor even that they can necessarily be generalised to Chinese or Koreans, but simply that candidates for non-genetic mechanisms are readier to hand than Lynn acknowledges.

One might also note the conclusion of a 1993 World Bank survey of East Asia, which said that heavy investment in education was one of the few policies successful countries had in common. But Lynn's general rejoinder to comments of this kind is to invoke the idea of gene-culture co-evolution. In other words, intelligent people create environments, whether at the level of the family or of the nation, that are conducive to the development of intelligence. Smart genes take care of themselves.

This applies to class hierarchies as well as racial ones. Like other hereditarians, Lynn believes that intelligence determines wealth rather than the other way around; he considers that social mobility has been extensive enough for long enough to have allowed clever genes to have concentrated themselves in the upper strata of society. He also sees connections between low intelligence and delinquency. In 1993, he appeared in a BBC *Horizon* documentary on the roots of violence. This was the occasion for a remarkable encounter with a youth called Mick Money, whose resemblance to a character in a Martin Amis novel did not end with his name. Money, who had been involved in petty violence, settled down with a cigarette in his hand, cans much in evidence around him, as Lynn asked questions aimed at assessing his intelligence.[16]

The critical question was about deferring gratification. If the Professor offered Mick £100 on the spot, or £200 in a year's time, which would he take? Mick opted for the cash up front. 'If I had to work, say, a week, and the geezer said "I'll give you £100 for the week, or you can have £200 a year later", are you gonna work all week and then trust the geezer?' Money asked Lynn. This seemed to demonstrate an ability to learn from experience and to make useful predictions about human behaviour. Sympathetic assessors might also consider Money to have shown verbal dexterity and social skills in his use of a hypothetical example which avoided any direct suggestion that Professor Lynn might not pay up.

For Lynn, however, this attitude was strong evidence that Mick Money was 'a typical psychopathic personality'. Money's own diagnosis was that Lynn failed to evaluate his theoretical models empirically: 'The geezer had it all sorted before he came here. He already had his picture in his mind . . .'

Lynn's readiness to venture onto the terrain of clinical psychology illustrates a general confidence about explanatory power which many psychologists seem to share. In particular, evolutionary history is a game everybody feels able to play. Its technical mechanisms, particularly game theory, are familiar in an age of systems theory and liberal economics. Like others, Lynn sees organic connections between Darwinism, social Darwinism and market forces. In *The Secret of the Miracle Economy, Different National Attitudes to Competitiveness and Money*, published by a conservative think-tank, Lynn invoked Hayek and Darwin in the same breath.[17] He observed that 'economics can be regarded as a branch of the more general biological theory of evolution: the branch dealing with the application to human economic affairs of the principle of competition leading to progressive improvement through the survival of the fittest.'

The phrasing reveals that Lynn is one of those who believes evolution is going somewhere, and that progress in its modern social sense can be related to progress in the sense of gradualistic adaptation. This is the ascent of Capitalist Man, through a competition which Lynn believes was decided in the last Ice Age.

As in many evolutionary accounts of human adaptation, temperature is the dominant factor. But instead of uncontroversially confining its influence to the shape of the nose or the length of limbs, in Lynn's account it selects for intelligence by promoting an increase in brain size and neural speed. Lynn builds on the racially marked version of the hunting hypothesis, in which Africa, Carleton Coon's 'indifferent kindergarten', is lax in stimulating hominid and modern human cognitive development. Africans ancient and modern had their meals to hand, provided by a benign maternalistic environment. In the Kalahari today, !Kung people continue to exist without knowing the meaning of hard work or competition, devoting only one day in three to acquiring food. 'The rest of the time can be spent relaxing about the camp,' Lynn remarks.[18]

There is no hint in Lynn's writing, as there is in late-period Coon, that such people might have the right idea after all. His concern is with the struggle between nations in the modern capitalist economy, not whether that structure might ultimately prove unstable.

The crucible of human progress, according to Lynn, lies somewhere in north-eastern Asia, near Lake Baikal. Here, he believes, the Mongoloids emerged some ten thousand years ago, shaped by the extreme cold of the last Ice Age.[19] As Coon proposed, their distinctive features can be seen as an adaptation to cold: their limbs shortened and faces flattened to conserve heat, the epicanthic fold of flesh at the corner of their eyes to protect against the cold and the dazzling snow. Lynn argues that changes took place inside the cranium as well. To survive in a cold environment, hunting skills need to be honed, and large game becomes increasingly important. The Mongoloids required good eyesight and visuospatial skills: survival might depend on the ability to spot an Arctic hare on a featureless icy waste, and to hit it with a missile; to get safely back to the group, an aptitude for orientation would be vital. The hunters also needed to be organised, developing strategies to co-ordinate hunts for large animals. Tactical skills were used to drive prey over precipices, while food-processing demanded a wide range of tools. Shelter needed to be found or built, clothes made, and fires lit.

Anticipating the objection that survival in Australia would also seem to require visual acuity and a keen sense of orientation, Lynn argues that the only major difficulty in the desert is finding water, food being available all year round, and in any case most Aborigines lived in river valleys until the Europeans drove them into the desert. But the ability of central Australian aborigines to conserve heat by vasoconstriction suggests long exposure to the desert environment: this is supported by the archaeological evidence, which indicates that the first Australians had occupied all the continent's major ecological zones by 30,000 years ago.[20] And if the rest have only been there for a couple of hundred years, the speed with which they got the hang of it suggests an impressive cognitive ability not reflected in their IQ scores – Lynn cites a figure from Queensland of 85. There have in fact been several reports that Aborigines have higher visual acuity, better spatial skills and visual memory than Caucasians. A complementary anatomical study has found the visual cortex to be larger in the brains of Aboriginal men than in those of Caucasians; the authors are at pains to stress that the visual system is extremely adaptable, and so environmental factors may underlie the difference.[21]

Findings like these lend weight to the sceptical view that evolutionary accounts are no more than just-so stories. It was the 'armchair adaptationists' who all but exhausted the credibility of Darwinian theory by the end of the nineteenth century.[22]

Nobody would dispute, however, that the prospects of survival in a cold climate are greatly improved by discussion and planning. With plant foods unavailable for much of the year, the ability to think ahead becomes important. This can then be applied to other problems, and facilitates an increasingly sophisticated social organisation. Christopher Stringer and Clive Gamble have suggested an enhanced capacity for 'planning depth' as a reason why modern humans outcompeted the Neanderthals in Europe. The newer arrivals appear to have been able to send teams to particular locations to take advantage of food sources. 'Such an elaborate planning system would work only if the returns from separate foraging parties were stored and shared at a future date,' Stringer and Gamble conclude.[23] The Neanderthals could only think in terms of the Upper Palaeolithic equivalent of £100 up front, whereas the Cro-Magnons and their relatives could grasp a concept equivalent to that of £200 in a year's time.

In line with the traditions of the hunting hypothesis, Lynn is mainly concerned with male intelligence. He notes the finding that men score two points higher on visuospatial IQ scales than women, reflecting a tendency which, he and other evolutionary thinkers suggest, began with the australopithecines, and which explains why boys today can throw stones better than girls. The hunting type is epitomised by 'strong silent males', as Lynn puts it. Like Clint Eastwood, they would be strong on visuospatial skills and short on words.

This produces a difficulty which Lynn skips over. If planning and co-ordination were important, hunters would need good verbal abilities as well. But Lynn's model of Mongoloid intelligence is based on extreme masculinisation. He suggests that the disparity between the visuospatial skills in Mongoloids and Caucasoids, at about eight points rather than two, 'mirrors but exaggerates' that between men and women. The underlying process is a trade-off of brain capacity, the implication being that visuospatial abilities were more important than verbal ones for the Mongoloid hunters.

Lynn then invokes a model of brain function which proposes that, in men, verbal processing is conducted exclusively by the left hemisphere, and visuospatial processing entirely by the right hemisphere. In women, by contrast, additional capacity is allocated to verbal processing in the right hemisphere. Thus damage to the left hemisphere impairs speech in men more than it does in women, who still have the secondary verbal faculty in the right hemisphere.

In the Japanese, Lynn goes on to suggest, the process of freeing space in the right hemisphere for visuospatial use has led to the transfer of various sensory odds and ends to the left hemisphere. His evidence comes from research which found that vowel sounds, animal noises, insect song and Japanese music are processed in the left hemispheres of the Japanese, but the right hemispheres of westerners. For the Japanese scientist who made these discoveries, however, the probable reason was the structure of the Japanese language. And if a cultural factor can produce a distinctive form of neural organisation like this, perhaps the visuospatial bias of Mongoloid intelligence might have a similar cause.

A more pressing internal problem for the Ice Age intelligence hypothesis is to be found across the Bering Strait, in the Americas. The indigenous peoples of the New World bear many physical resemblances to East Asians, and there is no doubt that the Americas were first populated by migrants who crossed the land bridge that used to join Alaska to the Asian land mass. Their descendants are classed as Mongoloids, but they do not reach the intellectual heights Lynn expects of that race. The median figure of the studies he cites is 89, a few points above the Negroid-Caucasoid hybrids, or African-Americans. However, he reports, they fit his model by doing better on visuospatial tests than verbal ones. His explanation is that the ice moulded the basic Mongoloid traits 40,000 years ago or more, but it took the final Würm glaciation, which lasted from 24,000 to 10,000 years ago, to forge the definitive Mongoloid character.

The scenario only holds water if the first settlers crossed the Bering bridge before the emergence of the 'classic' Mongoloid. Some archaeologists have indeed claimed discoveries, in California, Chile and Brazil, dating back 30,000 years or more. Others, however, argue that the migrants began to arrive around 15,000 years ago. The debate between the two camps is one of the fiercest controversies in archaeology. Reviewing the evidence, Maria T. Phelps, a student of Rushton's at the University of Western Ontario, finds Lynn's position unsupported. In her *Mankind Quarterly* paper, she argues that the intelligence of Amerindians was lowered by dysgenic influences. European diseases, which at home culled mainly the poor, attacked all strata of society in the New World, diminishing the intelligent elite. But the disproportionate mortality among the European poor resulted from the normal adjustment of parasite and host, whereby the two come to an accommodation: although mortality is reduced, the less healthy strata of society may remain vulnerable. In the initial stages, however, all sections of the population are taken by surprise. According to eugenic theory, the European elite would be expected to retain a dysgenic legacy from these early depredations.

More concretely, Phelps notes that the upper classes were also selectively exterminated by the Spanish invaders, who needed to decapitate the hierarchical Aztec and Inca societies in order to impose political control.[24]

That still leaves the North American Indians, though, and it also leaves the Inuit. Lynn acknowledges that the people formerly known as Eskimos pose one of the biggest obstacles to his theory.[25] They only arrived in Arctic America about 10,000 years ago; they are Mongoloid in form, and if anybody is cold-adapted, they are. Their visuospatial abilities are evident from their ability to live by hunting on the ice field, and there is also a body of text, from anthropological literature to fiction, which suggests that these aptitudes are generalisable. In his contribution to the *Current Anthropology* debate on scientific racism, Ashley Montagu quoted a report of the 'astonishing mechanical abilities' which Aivilik Eskimos of Southampton Island applied to the repair of engines or instruments, some of which American mechanics had abandoned in despair.[26] Thirty years later, Peter Høeg made the eponymous heroine of his novel *Miss Smilla's Feeling for Snow* a bicultural Greenlander with a gift for orientation and a passion for Euclid; the boy she befriends is mechanically adept.[27]

The problem for Richard Lynn is that his reading of the IQ data suggests that the Inuit are not very intelligent. One source, Philip E. Vernon, records a mean g of 91, and a mean visuospatial score of 90 (though he also notes a study in which Inuit boys' scores matched those of whites whose fathers had similar occupations).[28] This would come as no surprise to Montagu, whose point was precisely that intelligence tests do not cross cultures, and are therefore discriminatory. Nor would it surprise anybody who blames poor scores on oppression or social crisis. The first Southerners to educate Inuit children in Canada were missionaries, who banned native languages and were accused of brutality. In the 1950s, the nomadic way of life was finished off by starvation and government action, which included the shooting of sled dogs by the Mounties.[29]

But a theory that lives by IQ, dies by IQ. It is not surprising that the Inuit are reported to have good visuospatial skills, since they could hardly have got by in the Arctic without them. Nor is it unlikely that the extremity of conditions, insisting upon the need to make sense of a monochrome landscape and to rely upon animals for food, imposed strong selective pressures. Poor eyesight might well lessen the likelihood that an individual would survive long enough to reproduce. Higher-level cognitive skills, such as orientation, would also be at a premium – though that might be addressed by the formidable human capacity to learn from experience, rather than genetic selection. Whatever the processes underlying cognitive ability, however, a hypothesis based on IQ testing has to explain why the Chinese and Japanese score highly, although their homelands are no longer exceptionally cold, while the one population still living under the suggested selective pressures, the Inuit, scores poorly.

The evolutionary hypothesis runs into more trouble in Africa, when Lynn advances it to explain the curious manner in which agriculture appeared. Within the space of a few thousand years, people started farming in the Middle East, China and Central America. Orthodox commentators suggest that similar dynamics were at work around the world: an increase in population densities, which would tend to encourage cultural innovation; global climatic changes in the wake of the Ice Age, which may have altered the food resources available, and spurred people to develop new ways of securing their supplies.[30] Lynn, however, claims that the enabling condition was an increase in intelligence forced by the ice itself, which 'also explains why these advances were never made by the Negroids or the south east Asian races who escaped the rigors of the last glaciation'.[31] Yet Equatorial Africa was one of the first regions where farming was practised, and knowledge of cultivation was taken south of the Sahara region by Negroid people.[32] Agriculture was followed by the building of cities and the emergence of state societies in black Africa, just like everywhere else. Believers in white superiority may argue that Africans might have adopted agriculture, but they did not invent it. The same, however, could be said for Europeans. It is hard to imagine that Lynn's statement would have survived the peer review process in any journal other than the *Mankind Quarterly*.

Nor would his ideas about the evolution of intelligence find much favour outside hardline race science and its hinterland of hereditarian psychology. It is not just a matter of their racial dimensions, but their lack of connection with current thinking – except for selective borrowings such as those about how men and women process language. The idea that human intelligence was driven up to its current levels by males hunting and making tools has given ground to accounts based on social dynamics, which propose that human intelligence grew because there were advantages to having insight into the behaviour of other members of the group; to increasingly sophisticated communication within the group; and to having the wit to deceive one's companions. Groups are groups, whatever the temperature. It may be that, as far as intelligence is concerned, evolutionary pressure is other people.

If so, then Africa would be just as good as a kindergarten, in Coon's phrasing, as any other region of the planet. It might even be rather better, since the relatively benign environmental conditions would permit groups to grow in size and social complexity, in contrast to the stripped-down survival mode that might be forced upon pioneer bands in the north. The emerging discipline of evolutionary psychology is driven by the idea of sexual selection, whereby traits emerge as a result of competition between members of the same sex, or choice of mates by members of the opposite sex. One evolutionary psychologist, Geoffrey Miller, goes so far as to suggest that intelligence evolved as a courtship device.[33]

No matter what stimulating ideas emerge from contemporary evolutionary theory, however, scientists dedicated to the search for simple truth about human hierarchy seem unable to resist the allure of the simplest measure of all. We are more intelligent and have larger brains than apes, who are more intelligent and have larger brains than dogs, and so on down towards the flatworms. Fossil discoveries suggest the same trend exists in evolutionary time: *Homo sapiens* has a bigger brain than *Homo erectus*, who had a bigger brain than the australopithecines, and only one of us will leave silicon chips rather than stone choppers with our bones.

The idea that size means smartness appeals to intuition and informs popular myth, underlying the romantic countercultural notion of the 1960s that the intelligence of dolphins is similar to ours. Although their brains may be as large as human ones, their mental abilities appear to be roughly equivalent to those of dogs.[34] Within our own species, according to the South African anthropologist Phillip Tobias, 'it is likely that . . . few branches of human biology are more bedevilled with half truths, selective omissions, and the forthright proclamation of unproved propositions than that which deals with brain size and, especially, the relationship between brain size and function.'[35]

Taking the long view, Tobias and other workers have identified a curiosity in the fossil record. While the theme of thousands of millennia has been one of increasing brain size and increasing cognitive power, the upward trend in brain size has reversed during the last twenty thousand years, which covers most of the period in which human cultural development – as measured by technology and organisation – lifted off from a basically horizontal to a practically vertical trajectory. During the span of time that saw the painting of the Lascaux caves, the development of agriculture, and all the events of recorded history, average brain size has slightly diminished by 100–200cm^3. The drop appears to be associated with an overall decrease in body size and robustness. Among the suggested causes for this are the disappearance of big game, such as mammoths, after the Ice Age, which limited the scope for individuals to monopolise food resources, thus discouraging polygamy and reducing competition among men for women. Alternatively, humankind is argued to have shrunk – along with other species, such as wombats – as the globe warmed at the end of the Pleistocene.[36]

In craniology's nineteenth-century heyday, as Stephen Jay Gould has detailed, attempts to establish brain size as an index of mental capacity became more convoluted than the organ on which they were fixated.[37] Even at the level of the unrepresentative example they were compromised, since much attention was paid to the discovery that certain great men, notably Anatole France, had very small brains. Eventually, the enterprise was consigned to the attic of scientific folly, where the belief that value could be assigned by filling skulls with seed or shot languished for some time.

Then came Arthur Jensen, who discovered a 0.3 correlation between brain size and IQ. As noted earlier, IQ correlates positively with all sorts of odd variables, and which correlations are based on causal relationships is a matter of judgment. In Jensen's, this one was. Gould questioned why Jensen should declare a 0.3 correlation between IQ and brain size meaningful in this way, but deny any causal relationship for the 0.25 correlation between IQ and stature that he reported on the same page. A more credible explanation, Gould argued, is that tall people tend to be larger in general; and the correlation with IQ might be largely the result of poverty or poor nutrition, which can depress both IQ scores and stature.

By way of a riposte, Lynn cites the values Gould calculated for the data used by the nineteenth-century craniologist Samuel Morton, while J. Philippe Rushton cites a revisionist analysis of these data that denies Gould's charge of 'finagling'.[38] Gould's calculations were intended to compensate for this bias, which skewed the results towards white superiority and black inferiority. None the less, the corrected figures show a gap of four cubic inches between the African-derived and European-derived samples, the ratio between them being comparable to those found in later studies. These include a massive survey of nearly 20,000 skulls, which Lynn ranks according to race and size: Mongoloids are the largest, then Caucasoids, then Amerindians, South-East Asians, and finally Negroids.

The question is what this ranking actually means. According to the group who actually conducted the survey, cranial capacity is the result not of selection for intelligence, but of – once again – climatic adaptation.[39] They review various ideas about the relationship between brain size and other physical variables. One suggestion is that some metabolic factor governs brain size. Another possibility, advanced by Harry Jerison, is surface area. In Gould's words, 'brain weight is not regulated by body weight, but primarily by the body surfaces that serve as end points for so many innervations'.[40] The principle can be visualised by thinking of those decorative lamps made by allowing a sheaf of optic fibres to splay out into a hemispherical pattern: the more points of light at one end, the thicker the sheaf at the other.

Kenneth Beals and his colleagues found neither of these propositions entirely satisfactory. They calculated that body weight explains 39 per cent of variation in cranial size, surface area 38 per cent, and stature 6 per cent. When brain and body variables were compared with climate, brain size correlated more strongly, suggesting that it was more susceptible to climatic influence than the body as a whole. For populations in temperate and cold climates around the world, average cranial capacity was $1,386cm^3$, compared to $1,297cm^3$ for those living in hot climates. The Aleuts, an Arctic group closely related to the Inuit, have an average cranial capacity of $1,518cm^3$. Further south, the Choctaw Indians have an almost identical body surface area, but their average cranial capacity is $1,292cm^3$.

Beals and his colleagues base their argument on the fact that the ratio of surface area to volume is at its minimum in a sphere, which is therefore the ideal form for minimising heat loss. This allows them to explain both cranial capacity and brachycephalisation. As people moved northwards into colder regions, their heads became rounder and therefore bigger. Arctic peoples have short limbs, large heads, and minimal brow ridges, which disappear as the cranium expands above the eyes. Conversely, Australian Aborigines retain long slender limbs, long flattish heads, and pronounced brow ridges.

Cranial capacity is not an index of power like the engine capacity of a car, however. The brain is a soft wet object that occupies a rigid container. As the volume of the container increases, the brain may expand to fill the space available, but that does not necessarily mean that the number of neurons or the connections between them is increased. It may simply be a less compressed organ. There is some support for this view in a study which found that neurons are more densely packed in the cerebral cortices of women than in those of men, while they are less densely packed in larger brains of both sexes. Men's and women's brains contain about the same average number of neurons, although bigger brains in each sex do have more neurons than smaller ones.[41] In any case, despite all the celebrated advances of neuroscience, the study of the relationship between the physical structure of the brain and mental capacities is still in its kindergarten stage. For neuroscientists, exploring the microscopic intricacies of the most complex object we know, attributing significance to gross measures of size is frankly vulgar.

The psychometricians' rejoinder is that there is a positive correlation between brain size and IQ; and for a psychometrician, a positive correlation is always something to be cherished, no matter how small it may be. Lynn presents the results of sixteen surveys, dating back to 1906. In five of these, the correlation is 0.14; in five it is lower; and four are in the range between 0.18 and 0.23. Maciej Henneberg, who found one of the 0.14 correlations, pointed out that when correlations lie in the 0.1–0.2 range, less than 5 per cent of the variance in one characteristic is explained by the variance in the other one.[42] Noting that a 0.14 correlation explains 1.96 per cent of variance, Kenneth Beals observed in *Homo* that 'one might as well predict an IQ score by measuring a person's foot'.[43]

That still leaves two surveys, and both of these found correlations of 0.35. Lynn and Rushton have given particular weight to one of these, conducted by Lee Willerman (a fellow signatory of the 'Mainstream Science and Intelligence' statement), which assessed brain volume in living subjects by means of magnetic resonance imaging. There were only forty of these subjects, though, and as they were all students, they are not a representative sample of the population as a whole. Several other studies have been reported subsequently: Rushton cites correlations of 0.38, 0.40 and 0.43.[44] He does not, however, mention that the researchers who obtained the last of these figures found that correcting for age reduced it to 0.22, and further corrections for sex and head size made it all but vanish completely.[45] The others provide more authentic support for his views, though that is unsurprising in the case of the 0.40 figure, since it was generated by colleagues of his at the University of Western Ontario.[46] Doubtless many more data will emerge from the genre: the advent of brain scanning techniques seems certain to translate craniology from the nineteenth century to the twenty-first.

In the meantime, Rushton has amassed considerable data from

conventional sources to support his hypotheses about racial ranking. Aptly, his principal source of evidence is the military. In addition to an international survey of anthropometric data on servicemen, he analysed the figures gathered by the US Army to assess its needs for helmets of different sizes.[47] Excluding certain categories on grounds of ambiguity, such as warrant officers (not proper officers) or Hispanics (not a proper race), Rushton produced a pattern in which Mongoloids had bigger heads than Caucasoids, who had bigger heads than Negroids; men had bigger heads than women; and officers had bigger heads than other ranks. He noted, however, that the differences were not huge: the average Mongoloid head was only 4 per cent larger than the average Negroid head. Lee Willerman commented that even this difference only appeared because Rushton applied an inappropriate correction for body weight; Arthur Jensen made a similar point, adding that speed might be more important for intelligence than size.[48] Countering criticism that the military may not be typical of the general population, Rushton later supplemented his findings with ergonomic data on civilians from the International Labour Office in Geneva.[49]

Rushton has a keen sense of history and struggle. In his accounts, Gould's work is presented as a sort of high tide mark in the anti-racist scientific current that prevailed after the end of the Second World War. *The Mismeasure of Man*, he implies, represents the end of an era rather than the end of an argument. His own work is not the only manifestation of a new current, ascribing significance to the relationship between brain size and intelligence, which has arisen in the 1980s. On the other hand, he immodestly cites the opinion of a reviewer that his US Army figures 'may be the most valuable set of data in this whole literature'.

His arguments, and the related question of whether it is right to publish them in a scientific journal, occupied considerable space in the editorial and correspondence pages of *Nature* in 1992.[50] One contribution came from Christopher Brand, who argued that the racial difference in brain size was small, whereas the sexual one was quite large, and yet there is no difference in intelligence between the sexes.[51] This position, commanding consensus opinion among psychologists for many years, is the biggest obstacle for the hypothesis that Africans are less intelligent than other peoples, and that this is because they have smaller brains. It was not until 1994 that Richard Lynn finally published an analysis which managed to conclude that men are more intelligent than women after all.[52]

For Rushton, brain size is related not just to sex but to sexuality. In his rank ordering of racial traits, encompassing intelligence, maturation rates, personality, social organisation and sexuality, the latter is pivotal. The picture is a familiar one, since it conforms to long-standing stereotypes. Rushton would agree with the observation of *Le Japon 1994*, a text jointly published by the French Embassy in Tokyo and the Institut Franco-Japonais de Tokyo, that 'the Japanese tend towards inactivity in their sexual behaviour.' This, he argues, is a racial trait of Mongoloids, and it goes with the social discipline and emotional inexpressiveness characteristic of East Asian societies.

The Negroids, by contrast, he finds to be hypersexual by all measures, from the rate of dizygotic twinning to the likelihood of conception per copulation to the capacity of the black woman for orgasm. Kinsey's data provide rich pickings, enabling him to break down by race categories such as Number 329, 'Frequency of positions in coitus in first marriage: female above, male supine: "Much" '.[53]

Included among the parameters is the most mythologised one of all. Surveying the ethnographic literature, Rushton finds that the average dimensions of the erect Oriental penis are $4-5\frac{1}{2}$ inches in length, and 1.25 inches in diameter; of the Caucasian, $5\frac{1}{2}-6$ inches in length and 1.3 to 1.6 inches across; and of the black man, 6.25–8 inches in length, and 2 inches in diameter.[54]

Inches are not a common unit in contemporary scientific literature. Here they can be taken as a sign not only of the age of the source material, but also of the essentially populist character of the work. It is the Western male rather than the scientist who worries about this question, and in North America, he visualises it in inches. In Rushton's most recent account, however, metric measures are given in parentheses, allowing them to be compared with data relating to that contemporary preoccupation, the condom. He is able to report that 'both the World Health Organization's *Specifications and Guidelines for Condom Procurement* and the United Nations' International Organization for Standardization have recommended a 49mm flat width condom for Asia, a 52mm flat width for North America and Europe, and a 53mm size for Africa'.[55]

A millimetre here or there might not seem of overwhelming significance, but at least there is no arguing with the figures. The same cannot be said for some of his other sources on human sexuality, including his own research. Although no field of social science is more vulnerable to distortion by both modesty and boastfulness, Rushton is content to take what people say about their sexual behaviour at face value: accordingly, he distributed questionnaires to blacks, whites and East Asians at a local mall, asking them about their penis size and how far they could ejaculate.[56] The episode was never dramatised by the scriptwriters of *Beavis and Butt-head*, but it should have been.

Rushton also draws upon the work of an anonymous French Army surgeon, whose book *Untrodden Fields of Anthropology* was first published in 1896.[57] Rushton reacted indignantly when two of his critics described this work as 'anthroporn'. He protested that the author 'had spent 30 years as a specialist in genito-urinary diseases . . . He wrote the book on retirement and in a Preface hoped that one day the scientific world would be more enlightened than it then was about the study of sexuality.'[58] In fact, this sort of piety was typical of the genre pioneered by the book's publisher, Charles Carrington, the leading pornographer of the late nineteenth century. He styled himself a 'publisher of medical & scientific works', his speciality being titillatory material posing as scientific literature.

That much is evident from even a casual perusal of the book, which is essentially a guide to sexual tourism conducted by a voyeuristic racist. His travel tips include a warning that black women smell like crocodiles: 'This influence is particularly noticeable when she is excited by sexual passions, and is annoying to beginners who are not accustomed to it, but you end by getting used to it.'

Even here, there is ammunition for the critics who accuse Rushton of biased selection, since the picture painted in *Untrodden Fields of Anthropology* is somewhat more equivocal than Rushton's. The testicles of the Negro, remarks the surgeon, are often smaller than those of the European. The figures for the length of the erect Negro penis, measured in Guiana, are the same as those cited by Rushton for the race as a whole. 'But the erection is never hard like that of the European, the Chinese and the Hindoo,' the surgeon notes. 'It is always rather soft, and feels to the hand like a strong elastic hollow tube of black india-rubber.' The experimental method is not detailed.

The surgeon also claims that the testicles of Chinese men are only slightly smaller than those of Europeans. This is contradicted in an article written not by Rushton but by Jared Diamond, who cites findings that testicles in two Chinese samples are half the weight of those in a Danish sample.[59] In comparisons across species, he comments, large testicles correlate with high copulatory frequency and high likelihood that a female will mate with several males during an ovulatory cycle. In other words, as Roger Short has proposed, testis size is increased under conditions of 'sperm competition'. Indeed, as a later article in *Nature* notes, Caucasians appear to produce twice as many sperms a day as Chinese men.[60] But, *pace* Rushton, Diamond notes that 'an explicit test revealed no relation between testis size and copulatory frequency in Korean men'.

Like Rushton, however, he suggests the possibility of a link between testis size and the frequency of dizygotic twinning, which is taken as an indicator of a high ovulatory rate. The other authors, Paul Harvey and Robert May, consider it more likely that differences in testicular size between races are 'different responses to different mating behaviours'. In other words – though these are not spelled out – European men have larger testes than Chinese men because they are more sexually active.

John Maddox, *Nature*'s editor, argued in his leader on Rushton that 'unpalatable' scientific claims needed to be backed by 'extraordinarily compelling' evidence to justify publication, and that Rushton had failed to meet this requirement. The two articles on testis size that appeared in the same journal are an illustration of Rushton's significance. It lies in his ability to modernise nineteenth-century raciology, anthroporn and all, by weaving it into the discourse of contemporary evolutionary thinking. At one moment he appears marginal, if not mad; at the next, he is talking the language of a flourishing scientific mainstream. And thus Herrnstein and Murray feel able to say – albeit in one of their discreet footnotes – that his 'work is not that of a crackpot or a bigot'.[61]

Why, then, is he so isolated? Some of the reasons – the provocative statements, the dubious sources – are obvious. But beyond these, there lies his embarrassing capacity to demonstrate just how readily modern evolutionary ideas lend themselves to racialisation. There are several reasons why sociobiologists and evolutionary psychologists have avoided racial issues. One is that their project is to demonstrate the existence of a universal human nature, not to explore difference (other than that between males and females). Another is that the furore that greeted sociobiology in the 1970s has taught them the need for political discretion. A third is that they share the views about racial equality that prevail throughout society as a whole. But modern Darwinism appears to present few theoretical barriers to the racialisation that might result from a change in the *Zeitgeist*.

The purpose to which Rushton applies his eclectic collection of data is an evolutionary synthesis based on 'life history' theory. It revolves around a concept called r–K, developed in the 1960s by R. H. MacArthur and Edward O. Wilson (who later became famous, and controversial, as the leading theorist of sociobiology). Put crudely, an r selection strategy entails producing a lot of offspring and leaving them to their own devices, while a K strategy involves producing only a few offspring, but taking good care of them. The former is preferable when a species begins colonising a new habitat; then, as the species' numbers build up and approach K, the carrying capacity of the habitat, a strategy based on more efficient utilisation of resources is more effective. Two aspects of the theory are worth noting in this context. One is that r and K selection form a continuum, and a single population may shift from one to the other. The second is that K selection is favoured in environments that are predictable.

Whether it is appropriate to separate r and K from the other 37 factors which comprised the original model is open to question, as is its applicability to variation within the human species, but Rushton is not the only scholar to have applied r–K theory to people. Among them was the anthropologist Vernon Reynolds, who tried to apply the concept to human religions but eventually gave up.[62] Scientists with less quixotic interests have also lost their enthusiasm for r-K theory in recent years.[63]

The r-K continuum lends itself readily to a rank ordering of species, as Rushton illustrates by reproducing a diagram from Donald Johanson's successful popular book *Lucy: The Beginnings of Humankind*.[64] At the r end are oysters, which produce 500 million eggs a year, then come fish, frogs, rabbits, felines, and finally the great apes, which produce only one infant every five years or so. The Great Chain of Being is back, in the idiom of population biology. Rushton quotes Owen Lovejoy's summary: 'More brains, fewer eggs, more "K".' He also quotes E. O. Wilson to the effect that 'higher forms of social evolution should be favoured by K selection'. In a paper delivered to the 1989 American Association for the Advancement of Sciences annual conference, he observed that 'although it has become unfashionable to view man as the "most developed" of species, this once traditional view gains novel support from the perspective of an r-K dimension'. Or, in the crisper formulation of the synopsis, 'one theoretical possibility is that evolution is progressive and that some populations are more "advanced" than others'. In one discussion, Rushton speaks of 'more evolved' people, and goes so far as to suggest the restoration of that favourite nineteenth-century marker of racial rank, muzzle length (the trait to which Thomas Huxley alluded in his supercilious comment on 'our prognathous relative').[65]

Another of Rushton's tables lists the traits associated with each strategy. The r-strategist has many offspring, but does not look after them, and they have a high mortality rate. Members of an r-population mature early and start reproducing early. They are low in intelligence, altruism and social organisation. Members of more advanced K populations have the opposite characteristics. To explain why the Mongoloids should have gone further down the K road than the Negroids, and the Caucasoids should lie in between, he invokes the same mechanism as Richard Lynn: the evolutionary pressure of a cold climate. He favours the out-of-Africa model of modern human origins, though his theories are also compatible with multiregionalism.

Conspicuous by its absence from Rushton's reading of r-K theory is the principle that a population may shift its strategy in response to changes in circumstances. True to the traditions of race science, he sees racial characteristics as essentially invariant. Yet even a sociobiological notion like r-K takes on a radically different cast if the element of flexibility is retained. In Rushton's vision, black American men are behaving true to their race when they commit violent crimes and father children without taking responsibility for their care. Behaviour of this type is consistent with the fundamental principle of evolutionary psychology, that men and women have different reproductive interests and therefore pursue different strategies. Women will tend to select mates according to cues which suggest they will be good fathers, such as sound financial prospects or calmness of temperament. But at the same time, so the theory goes, they will be alert to cues which suggest that a man's genes are made of the right stuff, such as an aggressively competitive attitude towards other males. The optimal strategy, the evolutionary psychologists suggest, is to marry a citizen but to look out for the occasional stud with whom to cuckold him.[66]

When the prospects of reasonably paid work and job security are poor, the calm and caring man's potential to become a good father and provider will not be realisable. In the absence of the good, the baddest becomes the most attractive. We might speculate, with tongue not entirely in cheek, that one way to drive poor women into the arms of the toughest gangsters would be to implement Charles Murray's proposal to abolish welfare, the state's substitute for a breadwinner. If the element of flexibility is retained, an r-K interpretation makes observed behaviour in urban ghettoes look like a short-term adaptation to an unpredictable environment in which resources are scarce, rather than the expression of racial character.

Reactions to Rushton's theorising have been highly mixed. In a critique entitled 'Oysters, Rabbits and People', Marvin Zuckerman and Nathan Brody found one of Rushton's accounts 'flawed in terms of its obscure logic, selection of data, biased analyses of data, selective review of the literature, ignoring of large group difference within the major races . . . and aggregating that which should not be aggregated, lack of any examinations of possible interactions of race with social class . . . and a failure to assess statistical and scientific significance in terms of magnitudes of effects.' Using a more demotic idiom than is usual in academic journals, the sociobiologist David Barash dismissed Rushton's 'pious hope that by combining numerous little turds of variously tainted data, one can obtain a valuable result; but in fact, the outcome is merely a larger than average pile of shit'.[67]

Hardline race scientists like to think that their peers reject their work on political grounds. That is clearly one motivation, but another is embarrassment at what other scientists regard as very bad science. Rushton believes that the strength of his theory is its ability to incorporate a breathtakingly wide range of evidence, but that is precisely its greatest weakness. It proposes that everything in human life, from the fortunes of nations to the length of pregnancy in different ethnic groups, was ultimately determined by the temperature. Reductionism has its place in science, but this is ridiculous – unless one has a deep-seated belief in racial differences. Denied that context, Rushton's theories have little persuasive power.

In Rushton's world, race is the only variable, whether the data are social or biological. He adduces crime figures reported to Interpol as evidence that Negroids are more violent than Caucasoids and Mongoloids.[68] No matter that different countries have different histories, or that murder rates are generally high in countries where the machinery of the state is ineffective, such as Russia since the collapse of the Soviet Union; race is the only factor that Rushton feels necessary to extract from humankind's rich social tapestry. Furthermore, his argument assumes that the cultural distinction between murder and war expresses a biological truth: that violent behaviour conducted on individual initiative is different in biological character from that which is organised by the state.[69] But the biology of the Russians did not change between the era of Stalin's terror and the brutal chaos of the 1990s; their social and political structures did.

In his conviction that all evidence conforms to the pattern he describes, Rushton is prone to overlook anything that does not fit. One small example is first molar eruption. B. Holly Smith, a University of Michigan anthropologist, has calculated that in 21 primate species, the time that the first molar tooth erupts has been found to have a 0.98 correlation with brain size. In other words, the later the teeth emerge, the bigger the brain, and the relationship between the two could hardly be tighter. When Rushton read of these findings (in Leakey and Lewin's *Origins Reconsidered*), he checked Eveleth and Tanner's *Worldwide Variation in Growth & Development* to see whether there was a racial pattern within the human species.[70] The answer was 'not as clear as you might expect', but he was satisfied that the first permanent molar tooth erupts earlier in Africans than Caucasians and Asians. Here was another piece of evidence that 'fits too perfectly'.[71] In *Race, Evolution, and Behavior*, he gives the average ages at which children start to acquire their permanent teeth (beginning with the first molar) as 5.8 years in Africans, compared to 6.1 in Europeans and East Asians.

To achieve perfection, however, Rushton had to discard three of Eveleth and Tanner's seven 'Asiatic' sets of figures. From Rushton and Lynn's typological perspective, it might perhaps be valid to drop the Canadian Amerindians and the 'Mongolid' Indian tribe called the Khasi. But there can be no excuse for excluding the classically Mongoloid Eskimos. Once again, the awkward Inuit confound Rushton and Lynn's racial schema. Their permanent teeth were found to start erupting at 5.5 years, the second earliest age recorded anywhere in the world. They bring the average Asiatic age down to 6.0 – the same value is obtained when all the figures are included – and thus place Asiatics between Europeans and Africans. Underneath the perfection of the African average, moreover, lies an African-American figure of 6.2, exceeded only by the sample population's European neighbours in Michigan.[72]

These esoteric figures give weight to the criticism that, as Mick Money would put it, Rushton already has his picture in his mind. After all, his vision is so intense that he sees no difficulty in identifying race as a principle that unites children's molar teeth and the 'undulating rhythms' of African dance noted by the French Army Surgeon, which he sees as a tactic serving the r strategy.[73] It can be argued that scientific insight consists to a great extent in seeing beyond superficial impressions and patterns that seem obvious, but Rushton considers stereotypes to be valid evidence for his theories. 'The overwhelming majority of stereotypes seem to me to have more than a kernel of truth in them,' he declares. Find some data to support them, add some 'really unusual data like twinning rates that is compatible with it', and the environmentalist critic has a tough case to answer.[74] This is evolutionary populism, and it has a natural constituency among those who think racial prejudice is just common sense. Rushton's appearance on the Geraldo Rivera show might just be the beginning.

But is it science? Some scientists, such as the geneticist David Suzuki and the palaeoanthropologist C. Loring Brace, say it is not.[75] The answer may depend on whether you think it scientifically meaningful to conclude, as one study cited by Rushton did, that African-American youths have higher self-esteem than whites or Orientals by noting their responses to statements such as 'I feel I am a person of worth, on an equal basis with others', and whether you consider it valid to link this finding with the suggestion that African-Americans' self-esteem arises in part from their testosterone levels.[76]

This sort of scholarship tends to massage the self-esteem of physicists, chemists and biologists by reassuring them that psychology is not a science. However, it is probably more useful in this context to say that if the work looks something like science, then it should be regarded as science.

There are two basic reasons for taking this position. First, what counts is the institution and its power: psychologists are influential, regardless of what 'hard' scientists and sections of lay opinion may think of them. Second, Rushton and Lynn operate within scientific discourse, subjecting their papers to peer criticism, and – most importantly – generating testable hypotheses from at least part of their work.

None the less, their support base remains limited. While Rushton has drawn upon a range of sources, from the hot tub end of social psychology through population biology to neuroscience, he has had very little success in building alliances in any of these areas. His failure to gain endorsement from neuroscientists for his ideas about brain size is particularly telling. He has his loyal supporters, but the names lined up on the jacket to endorse *Race, Evolution, and Behavior* are a familiar coterie of committed inegalitarians, including Jensen, Eysenck, Lynn, Michael Levin and Thomas Bouchard.

Although the neo-Galtonian project as a whole is expanding

relentlessly, Rushton's contribution to it has not yet stimulated much research by other workers – in contrast to the reams of critical responses it has provoked from his opponents. One exception is Lee Ellis, who has enthusiastically adopted r-K theory to explain black criminality.[77] His distinctive contribution to the field is the suggestion that black men are disposed to commit rape in pursuit of the r strategy.

One of the main theoretical objections to the Rushtonian idea of r, K and sex arises from the observation that women's sexual behaviour is not confined to a short fertile period, as in other species. The currently favoured explanation is that continuous sex offers a reward to males for pair-bonding, thus underpinning the family. If that is the case, critics have pointed out, the most K populations should be the most highly sexed.[78]

Another objector is Richard Lynn, despite the views he and Rushton hold in common about brain size and the Ice Ages. The problem is the theory's prediction that K selection is favoured in predictable environments. Lynn can see no reason why the cold north should be any more predictable than the tropics.[79] In order to justify this idea, Rushton argues that the Arctic was harsh but predictable – 'people knew that it would be difficult to find food for 4 to 6 months every year' – whereas on the African savanna, people were vulnerable to unpredictable droughts and epidemics.[80] This is a radically different picture from the traditional one of the 'indifferent kindergarten', where people supposedly found life's necessities at their fingertips. Since mental agility is so often the key to coping with the unpredictable, and modern Africans encompass more genetic variation than anybody else, the savanna might be expected to pick out any genes that enhanced cognitive powers.

Such a possiblity has been acknowledged by Edward Miller, a professor of economics, in a critique that stands Rushton's theory on its head.[81] Miller argues that drought ought to favour a K strategy, since K is geared to making the most of limited resources. It might also, he acknowledges, favour intelligence and social discipline. On paper, he concludes, the environment in which Negroids evolved should make them the most K-selected race. But he is fully persuaded by the inventory of traits produced by Lynn, Rushton, Ellis and others. These must therefore be the result of a different mechanism, and his suggestion is 'paternal provisioning'. Women in hot regions could find enough food for themselves and their children, whereas in colder latitudes the women had to rely on men to bring back food from the hunt. They therefore selected mates on the basis of their apparent potential as providers, rather than for immediate sexual gratification. Meanwhile in the tropics, men were free to spend their time pursuing new sexual opportunities and fighting with other men. Miller thus identifies evolutionary mechanisms for the muscular build observed in black males, and for the sexual relationships between black women and men.

Perhaps the most remarkable discussion of the savanna and African evolution does not refer to Rushton or to his ilk at all.[82] Throughout her account, Fatimah Jackson emphasises the tremendous variability of Africans. They were formed not just by the savanna but by desert, rainforest and steppe as well. In the New World, Jackson argues, they were biologically shaped by white oppression. One of their adaptive responses was the maintenance of a cluster of traits that shorten the time taken to reach sexual maturity, prolong the period of fertility, decrease the duration of pregnancy, and reduce the period of infantile dependency. These adjustments include early puberty, narrow pelvic breadth in women, combined with a smaller head size in babies at the time of birth, accelerated foetal development, and an increased rate of dizygotic twinning. Together, they should add up to an enhancement of reproductive potential.

The roots of this adaptive package, Jackson argues, probably lie in the tropical savanna and forest. Tropical habitats are rich in species, but the numbers of each species tend to be low. The ability to accelerate reproduction and variability is advantageous under such conditions; and Jackson argues that it continued to be advantageous under the conditions of oppression imposed on Africans in the New World. 'Many African-American philosophers have written cogently about their perception of the group as being "under siege",' she observes, 'and this recognition of the continuing difficulties that African-Americans face within a system dominated by persons of European descent cannot be decoupled from the concept of reproduction as an adaptive stress response to an adverse, density restrictive environment.'

Jackson goes on to suggest that the precocious development of black children may have its origins in the savanna, the ideal environment to encourage walking. In other environments, where steep slopes or exposure to cold were more likely to await the inexperienced child, learning to walk may have been less advantageous. Studies of the onset of bipedality in peoples associated with different latitudes might prove informative on this score, Jackson observes. An original propensity for accelerated development – including higher bone mineralisation rates and greater visual acuity (*pace* Lynn) – might then have resonated with selection pressures under slavery, when physical capacities were at a premium.

Among the 'biologically significant adaptive domains' identified by Jackson as being under social and political influence in African-Americans are 'intellectual ability' and 'spirituality'. Jackson's writing, here and elsewhere, is imbued with a strong sense of commitment to the African-American community of which she is part. At the social and political level, it's hard to imagine two scientists with more distance between them than Jackson and Rushton.

Hitherto, anti-racists and those who could be loosely sorted to the left half of the political spectrum have tended to reject the association of biological and social dimensions of human life. Even if they accepted the principle of interactions between genes and culture in principle – and all the scientists among them did, despite the *canards* to the contrary put about by hereditarians – they tended to deny the practical possibility of investigating such relationships. Yet here is an anti-racist, pro-African-American scientist talking about evolution, intelligence and spirituality in almost the same breath – and in the *Journal of Black Studies*, which promotes the Afrocentric worldview of its editor, Molefi Kete Asante. She does not dismiss the significance of phenomena such as dizygotic twinning rates, but agrees that these reflect evolutionarily shaped reproductive strategies – which are currently expressed in response to social and political oppression. Her genetics are interwoven with Africanist elements; a reference to a model of ethnicity proposed by the Senegalese scholar Cheikh Anta

Diop in one paper, diagrams based on traditional African designs in another.[83]

Rushton's work may well turn out to be part of an 'emerging paradigm', as he hopes, but a neo-Darwinian paradigm is not the same thing as the neo-Galtonian worldview he espouses.[84] Jackson's Afrocentrically inflected reading of natural selection is a striking illustration of the ideological diversity that can flourish under Darwinian skies. Yet 'can' is a long way from 'will'. The extent of diversity within Darwinism will depend upon the extent to which scientists in general are persuaded that the study of 'gene-culture co-evolution' is a viable scientific pursuit, and that its social benefits outweigh its dangers.

The peculiar contribution of Rushton and Lynn to race science is to have proposed a racial hierarchy which does not place their own racial group at the top. Theirs is post-colonial race science, in which anxious white perceptions of the current state of play in the struggle between nations are translated into a theory considered to hold true for the entire course of human evolution. On the eve of the millennium and the Pacific century, the innate superiority of the Oriental is seen to be re-asserting itself, after an anomalous five hundred years of European dominance. The plight of Africa, by contrast, is seen as the inevitable result of a biological inadequacy that has become acutely apparent in a world increasingly governed by cognitive ability. The white man's burden having been shed, however, the North is under no obligations to help ameliorate the African crisis.

Lynn and Rushton differ in their approaches to the sociopolitical implications of their work. Rushton adopts the posture of the disinterested scientist, disavowing any interest in policy. Lynn, on the other hand, is happy to talk about his belief in eugenics. He also accepts the label of 'racist', or 'scientific racist', in the sense of 'one who believes there are race differences in various characteristics, particularly things like intelligence and crime.'[85]

Although Rushton's views are in this respect indistinguishable from Lynn's, Rushton vigorously rejects the term.[86] 'People are always saying, "Oh, you say whites are superior to blacks," ' he told Adam Miller in an interview quoted in *Rolling Stone*. 'Even if you take something like athletic ability or sexuality – not to reinforce stereotypes or some such thing – but, you know, it's a trade-off: more brain or more penis. You can't have everything.'[87]

The difference in attitude between the two psychologists may arise in part from their different experiences in the public domain. Lynn has led a quiet life at the University of Ulster, largely untroubled by protest, apart from the odd student picket. He can be billed, as on the *Horizon* programme, as a world expert on intelligence, rather than as a controversial race scientist. His public interventions, such as a piece in *The Times* pegged to *The Bell Curve* and headlined 'Is man breeding himself back to the age of the apes?', may be the more influential for their author's discreet public profile.[88]

Ever since Rushton's address to the American Association for the Advancement of Science in 1989, however, his theories have been relegated to a subordinate place in a debate which has centred on him as an individual, on his motivation, and on his right to promulgate his views. A sustained campaign of protest led the University of Western Ontario to propose that he be permitted to teach only through the medium of videotape. Ontario's Premier, David Peterson, who publicly advocated Rushton's dismissal, called his work 'highly questionable, destructive, and offensive to the way Ontario thinks'. The Ontario police investigated Rushton under a statute directed against 'every one who . . . willfully promotes hatred against any identifiable group'. After consulting academic opinion, they concluded that, in the words of Attorney-General Ian Scott, Rushton's theories were 'loony, but not criminal'.[89]

In some ways, Rushton resembles Hans Eysenck, whose journal *Personality and Individual Differences* has provided an important platform for Rushton and Lynn. There are probably not many people who would be prepared to accept the entire landscape of Eysenck's maverick vision, which extends from a passionate belief in IQ tests and the psychometric ordering of personality traits through a sympathy with Jensenism to a rejection of medical orthodoxy on smoking and a 'conditional belief in parapsychological phenomena'.[90] But he has sufficient purchase on the mainstream to exert considerable influence on both psychological and public discourse.

Whereas Eysenck is as well known for his popular IQ books as for his controversial views, Rushton is publicly synonymous with race, sex and nothing else. None the less, he speaks in the contemporary idiom – of sociobiology and economics entwined, of competition and hormones. His ideas may gradually percolate into the periphery of the public imagination by indirect routes, such as the downpage newspaper article on Edward Miller's theory that appeared under the jocular headline 'Sex secrets of the caveman Casanovas'.[91]

In any case, as primatology encourages the public to embrace apes while privileged whites grow increasingly disinclined to embrace their fellow humans of other races, lay readers will draw their own conclusions from what they learn of contemporary scientific discourse. Reflecting upon evolution and the 'missing link', a columnist in the London *Sunday Telegraph* reported that orang-utans, chimpanzees and gorillas have been found to have IQs of 80–95. The following week, a letter to the editor pointed out that, going by the paper's recent front-page report on *The Bell Curve*, this 'would make them on a par mentally with American blacks'.[92] But that, in his view, was not sufficient reason to accord them rights.

7

Cave Men with Attitude

BESIDES INTELLIGENCE QUOTIENTS and reaction times, there is a third leg to Richard Lynn's 'global perspective' on race differences in intelligence, as set out in the *Mankind Quarterly*.[1] To give it a framework, he reproduces an inventory devised by John R. Baker as a means of assessing whether a race has attained civilisation.[2] If the subgroups of the race can generally tick off all or nearly all of the 21 items on the list, the race can be considered civilised.

The first is that people normally cover their genitals and the upper parts of their bodies with clothes. Other stipulations include personal hygiene, towns, money, a calendar, and the wheel. Civilised societies should also permit their members to 'go about their various concerns in times of peace without danger of attack or arbitrary arrest', and eschew the use of torture. Finally, there must be 'some appreciation of the fine arts', and knowledge must be valued as an end in itself.

Reviewing African achievement, principally through the eyes of European explorers such as David Livingstone and the seemingly ubiquitous Francis Galton, Baker damns the 'Negrids' with faint praise. Wherever Africans demonstrate skill, outside influence is suspected. In most cases this is assumed to be Arab, but Baker admits that an Islamic culture would hardly be likely to have encouraged the Ife of Benin to make bronze images of humans. Perhaps the wide-ranging Greeks were responsible, he suggests, also noting that some of the heads appeared to have certain Europid features.

Another psychologist to have endorsed Baker's views on civilisation is Hans Eysenck, in an interview given to a fringe far-right magazine in 1977.[3] This style of argument, which allows Richard

Lynn to conclude that the Negroids 'achieved virtually none of the criteria of civilization', does not feature in Eysenck's *Personality and Individual Differences* or in the other academically accepted journals which have published papers by Rushton and Lynn. The line is crossed in the *Mankind Quarterly* and in Rushton's book *Race, Evolution, and Behavior*, which reproduces both Baker's list and Lynn's claim that the Negroids never made the Neolithic transition. This pattern of editing may make the civilisation argument look like an extreme reached only by those willing to take the racial hypothesis beyond the realms of anything recognisable as science. Rather than being the destination of the racial discourse, however, it may be the point of departure, only acknowledged publicly by the most indiscreet of true believers. As such, it helps explain the bizarre character of its Afrocentric mirror image.

In its broadest sense, Afrocentrism refers to a consciousness of the African origins of New World black people, an embracing of African cultural symbols, an assertion of kinship, and a belief that European bias has impeded the understanding of Africans and their culture. African-American society is nowadays permeated by a dilute Afrocentric consciousness. In its non-threatening variants, notably the vision of Bob Marley, which is centred in Africa but extends to humankind in general, the idea has even secured a foothold in the white imagination. But it also provides an ethnic creation myth, like those spun by the Romantic nationalists of Europe in the nineteenth century, which serves as the organising principle of separatist feeling. As whites withdraw much of the goodwill that Dr King and his comrades had persuaded them to extend, blacks become increasingly inclined to favour withdrawal from white-dominated society. The more florid and strident forms of Afrocentrism offer the possibility not just of living apart, but of conceiving the world according to fundamentally different concepts.

Afrocentrism's principle inspiration is the work of the Senegalese physicist Cheikh Anta Diop, who saw the erosion of the race concept as an ideological assault on Africans. Reasoning from a physiological principle called the 'Gloger Law', which ordained that mammals in hot and humid climates must be pigmented, he declared the fossil of a human who died in Africa 150,000 years ago to be Negroid. The first

Homo sapiens in Europe was a 'migrant Negroid' – represented by two fossils identified as such by European anthropologists at the turn of the century. The counterpart of 'Grimaldi Man', as the supposed type was known, was the Chancelade type. Since the latter remains resembled modern Eskimos, it was conjectured that the living population were descended from reindeer hunters who had followed their prey north as the ice retreated. As anthropologists came to find such confabulations implausible, the Grimaldi and Chancelade fossils were reconceived as examples of early European diversity, rather than racial types. Diop regarded the revisionists as 'Western ideologues' who sought to deny the racial truth of Negroid primacy.[4]

This adherence to outmoded ideas is a characteristic feature of Afrocentrism, a peculiarly Victorian ideology. Being a mirror image of traditional race science, it does its best to uphold the race concept in the face of contemporary academic scepticism.[5] A similarly fusty set of values is apparent in Afrocentric attitudes to arts and culture, such as the insistence of Molefi Kete Asante, Afrocentrism's most influential contemporary theorist, on describing Afro-American jazz as African 'classical' music.[6]

Diop asserted a privileged position for African precedence in all accounts of human advance, biological and cultural. His conviction that 'Ancient Egypt was a Negro civilisation' was shared by his compatriot Léopold Senghor, exponent of the concept of *négritude* and President of Senegal for seventeen years. Just as Europeans had taken all of black Africa, if not all of Africa, as an undifferentiated dark continent, Afrocentrists conflate African's great physical and cultural diversity into a single racial whole. The grandiose presumption that all Africans share the same values and worldview has been dubbed 'unanimism' by one of its critics, the Beninois philosopher Paulin Hountondji.[7]

Afrocentrism is metonymic: the part is taken for the whole. The identifiably black component of the Ancient Egyptian population is taken for all of it; Egypt stands for all of Africa. The Egyptian contribution to Greek civilisation is taken to be its essence, thanks in part to Martin Bernal and his book *Black Athena*. Yet, in fact, the reality is one of diversity. Bernal showed how nineteenth-century

European historians had developed a tradition of denying both blackness and civilisation in Ancient Egypt. But, as the exhibits accumulated by the eminently Victorian British Museum indicate, a variety of peoples lived on the banks of the Nile. Bernal believes that 'Egyptian civilization was fundamentally African', and that 'many of the most powerful dynasties which were based in Upper Egypt – the 1st, 11th, 12th and 18th – were made up of pharaohs whom one can usefully call black'.[8]

The pharaohs themselves would have been unlikely to see any use in the label, however. According to a summary of scholarly thought given by the anthropologist Bernard Ortiz de Montellano, Ancient Egyptians looked much like modern ones, becoming darker towards the south. They constituted a multiracial society which regarded itself as superior to others, but did not discriminate internally on grounds of colour. While sympathising with Bernal's central argument about the racist denial of Egyptian influence upon Greece, C. Loring Brace and his colleagues conclude, on the basis of their analyses of faces and skulls – and of Brace's vehement opposition to the race concept – that the 'attempt to force the Egyptians into either a "black" or a "white" category has no biological justification'.[9]

The flaw in Bernal's title – which would have been avoided by the choice of an alternative suggestion, *Egyptian Athena*, which Bernal has said he preferred – is that it encourages essentialism rather than the appreciation of diversity. On the T-shirts, too, the Egyptians are exclusively 'African' in appearance. They follow the line laid down by Diop, that 'the Egyptians were blacks of the same species as all natives of tropical Africa'.[10] The reality, as always in matters of human diversity, was more complex.

Afrocentrism in its more extreme formulations has two overlapping power bases, the youth and the educational system. One of the most widely publicised examples of the latter are the Portland African-American Baseline Essays, introduced in 1987 as required reading for junior school teachers in Portland, Oregon, and adopted in several other cities. One of these was Detroit, where Bernard Ortiz de Montellano chanced upon a copy while waiting for the start of a meeting about multicultural education. This is a cause he favours, as might be expected of a Mexican-American scholar whose publications include a book called *Aztec Medicine, Nutrition and Health*. On glancing through the science essay, he 'went absolutely ballistic'.[11]

Its thesis was that Africans are the 'wellspring of creativity and knowledge on which the foundation of all science, technology and engineering rests'. Among the exhibits presented in support of the claim is a small wooden effigy of the Egyptian deity Horus, which is argued to represent not a falcon-god but an aeroplane. This object also features in Ivan Van Sertima's collection *Blacks in Science, Ancient and Modern*, where it is presented as an example of 'African experimental aeronautics'.[12] Although it lacks a tailplane – presumed lost – and would have to be made of balsa rather than sycamore to fly, it convinced the author of the Portland essay, Hunter Havelin Adams III, that the ancient Egyptians had full-size gliders which they used for 'travel, expeditions and recreation'.[13]

Among the other wonders of ancient African science described in the essay is the Egyptian pregnancy test which involved pouring a sample of a woman's urine over barley grains: if they failed to grow, the result was negative. 'Modern experiments show this method was effective in about 40 per cent of tested cases,' Adams solemnly informs his readers. Improvements in efficiency presumably had to wait until the development of money, which would allow 50 per cent success rates to be attained by tossing coins.

Ancient Egyptians did not merely anticipate the conventional technologies of the twentieth century, however. They were 'known the world over as the masters of "magic" (psi): precognition, psychokinesis, remote viewing and other undeveloped human capabilities . . . Psi, as a true scientific discipline, is being seriously investigated at prestigious universities throughout the world.'

These declarations, accompanied by the dropping of bombastic terms such as 'psychoenergetics' and 'psychotronics', have the familiar piping tone of contemporary pseudoscience in hot pursuit of legitimacy. Pseudoscience has a radically contradictory attitude towards the real thing. On the one hand, it decks itself out with elements that mimic orthodox scientific practice, particularly jargon. The Portland essays also claimed orthodox credentials for Adams as a 'research scientist at Argonne National Laboratory'. It turned out that he was an industrial hygiene technician, with no educational qualifications beyond a high school diploma.

On the other hand, pseudoscience rejects the scientific paradigm and proclaims the imminent advent of a higher one, usually based on unspecified 'energies'. At the same time, it denies the claims of science to a special authority. In this cause, it finds a natural ally in cultural relativism. The recognition that one culture may not assume it has the measure of another, and that the concept of cultural hierarchy is broadly meaningless, is one of the century's greatest insights. It derives in large part from the work of Franz Boas, who did so much to destabilise race science by showing that skull form can be susceptible to environmental influence.

These days, though, Boasian precepts are apt to be enacted as farce. Ortiz de Montellano has reported a warning to teachers from the National Council for Social Studies, which is concerned that there is a conflict between the creation myths of Native American peoples and the hypothesis that the Americas were populated by migrants from Asia crossing the land bridge. The Lakota Indians, for example, want to have nothing to do with science. They reject the theory of evolution and believe they were created at a site called the Wind Cave, in the Black Hills of South Dakota. All they want from science is the remains of their ancestors, stored in New World race galleries, where they were accorded no more respect by the anthropologists who collected them than stones or plants.[14]

American museums have recognised the justice of this demand, made by Indian peoples as a whole, and the Lakota have the right to reject the scientific paradigm. But that is no reason for science to follow suit, as the National Council for Social Studies advised. 'The Bering Straits migration theory should be treated with great scepticism,' it warned, 'since there is absolutely no evidence (except logic) to support it.'[15]

In the world of advanced cultural relativism – a mish-mash of Romantic fantasy and expedient responses to political conflicts between interest groups – scepticism is directed at science and the rationalists of the *Skeptical Inquirer*, but not at the patent absurdities of Afrocentrism. This is also part of a wider phenomenon. People mistrust vaccines for their children, but rejoice in the cult of homoeopathic remedies containing nothing but water; they doubt that men landed on the moon, but are quite convinced that aliens have landed on Earth. This is in large part due to scientific illiteracy, but it is also the result of the association of science with exploitation, and with a social system that has estranged people from authentic values. Science, in short, is seen to be bad.

Africa, the birthplace of humanity, is felt to be a possible source of knowledge, presently lost, about how people should live. This belief is nurtured by Afrocentrist claims about African science, which in the end have nowhere else to go.

It is true that the gaze of Livingstone, Galton and many others was blind to much African cultural achievement, and was determined to ascribe what it did notice to outside influences, but there is no point in trying to maintain on the basis of the evidence that, say, African buildings or counting systems were as sophisticated as those of Asia or Europe. As a result, Afrocentrists resort either to fantasies about black Egyptians going up for a spin in their gliders on the weekend, or to the contention that Africans had access through their systems of knowledge to realms of reality unrecognised by Western science. Taking his cue from Diop, F. P. A. Oyedipe makes portentous reference to the 'New Physics' in his contribution to a collection of papers about Nigerian culture; and recommends the use of Kirlian photography, claimed to capture mysterious energies, to investigate 'Yoruba metaphysical aspects of healing'.[16] Most of the book, however, is devoted to accounts of prosaic technologies associated with activities such as food processing and textile production.

Afrocentrism tacitly recognises that it cannot win a competition based on comparisons of technology, which in effect entail acceptance of Western cultural values. That is not, however, what the conflict is really about. It is encapsulated in the race scientists' contention that individual violence perpetrated by black males is a sign of their lack of intelligence or foresight, whereas the collective violence perpetrated by white armies is a sign of their sophistication. Yehudi Webster, a sociologist, points out the basic dynamic underlying reactive Afrocentrism: the more strongly whites assert the mental inferiority of blacks, the more strongly blacks assert the moral superiority of black people.[17]

This fits neatly with the contemporary hostility to science, linking the perceptions that science is white and science is bad. It also speaks to a far older and broader opposition between black and white, reversing the association between whiteness and good on the one hand, and blackness and evil on the other. The power of Afrocentrist science lies in its symbolic and mythical dimensions, while its separation from orthodox scientific discourse allows it to be easily combined with other Afrocentrist, black nationalist and Black Muslim mythical elements. Combining the themes of conspiracy, the evilness of white science, and biological hierarchy, Frances Cress Welsing, a psychiatrist, claims that white fear of 'genetic annihilation' by superior blacks led to the invention of AIDS for the purpose of genocide.[18]

The most versatile ingredient of Afrocentric science is melanin, the compound which darkens skin. It also occurs in the brain, where its function is unclear. For the 'melanin scholars', however, it is black people's Kryptonite.[19] It is said to be a superconductor, to convert sound to light reversibly, to process information like a computer. People rich in melanin are said to have better co-ordination and so are better athletes, have extra-sensory perception, and be sensitive to other people's magnetic fields. They have evolved further than whites, possessing an 'essential melanic system' as well as a central nervous system. Their Achilles heel, however, is an affinity between cocaine and melanin that leaves black people vulnerable to the conspiracy to push drugs into the ghetto. Details of 'the latest uncoveries [sic] related to Melanin' are available via MelaNet, an Afrocentric region of cyberspace, as are a range of Afrocentric fashion items.[20]

The idea of an essential melanic system is an example of compensatory fantasy asserting black superiority in a dimension which white methods of inquiry, in this case IQ tests, cannot comprehend. According to a book by a psychologist called Nsenga Warfield-Coppeck, however, melanin should permit blacks 'to speak and read faster because they are pulling the basic patterns of how to do that out of their unconscious'. The compound can be linked not just to mental but social difference. 'In northern or European culture we see nuclear families, individualism and linear relationships,' says Clarence Glover of the Southern Methodist

University in Texas, who talks of melanin's 'electromagnetic pull' drawing blacks to light and music. 'In the tropical culture, there is a collective concept of "us and them", extended families and a respect for nature.'[21] Richard King, librarian of the MelaNet, argues that the loss of such faculties was the price Europeans had to pay for the light skins they needed in order to synthesise vitamin D in the gloomy North. By contrast, the Twa, a 'Pygmy' group, represent a *Homo erectus* population which was 'able to have direct communication with the angelic forms'. (King's source for this claim is a work published in 1913 by Albert Churchward, a retired Colonel of the Bengal Lancers who believed in the existence of Atlantis and Mu.)[22]

Melanin is also said to underlie black scientific achievement, from Egypt onwards. According to Frances Cress Welsing, the Dogon people of Mali knew that Sirius had a small faint companion star because the melanin in their pineal glands enabled them to sense it. All earthly events, furthermore, are converted to energy which beams up to Sirius B, where it remains accessible to black people via the melanin medium. She also maintains that George Washington Carver, who was born into slavery but became a distinguished agricultural chemist, knew what substances to extract from plants because they 'talked to his melanin and told him what they were good for' as he walked in the woods. Ortiz de Montellano points out that black schoolchildren are likely to infer from this tale that they do not need to study, just tune in to their melanin.[23]

As a scholar involved in multicultural education programmes, Montellano's main concern is that black children will be further disadvantaged by imbibing the egregious nonsense of extreme Afrocentrism. But the influence of the 'melanin scholars' and their colleagues is not confined to the poor and uneducated. In their entertaining polemic *Higher Superstition: The Academic Left and its Quarrels with Science*, Paul R. Gross and Norman Levitt quote praise for Ivan Van Sertima from the highly regarded African-American feminist scholar bell hooks [sic], who accepts his contention that Africans reached the Americas before Columbus.

Signs of this encounter, which hooks conceives as one in which the participants did not make 'a site of imperialist/cultural domination' out of their difference, are the thick 'African' lips of Olmec sculpture.[24] Just as John R. Baker saw traces of the European in Benin bronzes, and thus reassured himself that the impressive artistic achievement of the sculptors was not a truly African phenomenon, Afrocentrists take the physiognomy of Meso-American statues as evidence that the first American civilisations were influenced by Africa. On the surface, the message may appear to be that of the benefits of cultural exchange, but the underlying purpose is to assert that, everywhere one cares to look, Africa came first. The archaic preoccupations with the ancient world and with the notion of civilisation are the mirror image of traditional white racial scholarship; the methods are frequently identical.[25] Yet an acutely contemporary figure like hooks finds no difficulty integrating Afrocentrism into her discourse.

By contrast, Ortiz de Montellano was turned down when he and colleagues proposed a session on 'Multiculturalism and Pseudoanthropology' at a meeting of the American Anthropological Association. He says that he has encountered similar reluctance to engage with Afrocentrist pseudoscience among other scientific bodies. For Gross and Levitt, this confirms the 'decadence of a subject' – cultural anthropology – 'that was, at its height, not only scientifically important but intellectually bold and morally brave'.

Standing up to extreme Afrocentrists on the battlefield of campus politics may demand considerable moral bravery. As Yehudi Webster points out, it is not a falsifiable doctrine, since any criticism is declared invalid. 'I don't speak in class, because I don't want to get in trouble,' said a Haitian student who took Leonard Jeffries' course on 'World Civilizations' at City College, New York. 'If I try to say anything, they will say, "You are black on the outside, but white on the inside." '[26]

Afrocentrism's circularity is inherently authoritarian, and Leonard Jeffries has shown how to build that tendency into full-blown demagogy. In 1993, Jeffries won $400,000 in damages against City College (an institution with the dubious academic distinction of having given tenured posts to both him and his polar opposite Michael Levin, the philosophy professor with a passion for the farthest reaches of racial psychology) for removing him as chairman of the Black Studies Department. The occasion for the college's action had been a speech he made in 1991, in which he blamed Jews for the slave trade, and accused their descendants (along with Italians) in Hollywood of conspiring to cause 'the destruction of black people'.[27]

While the case was pending, Jeffries conducted a long campaign to oust the replacement chairman, Edmund Gordon. Among the 'ten good reasons why Edmund Gordon should resign', set out on a flyer issued by Jeffries's supporters, was the comment that Gordon is 'interracially married'. James Traub, writing in the *New Yorker*, describes the cultish entourage that surrounds Jeffries on campus. He and his students greet each other with the Egyptian salutation *Hautep*. They are also prepared to ask such spontaneous questions as the follow-up to Jeffries's reminiscences of a sojourn in Europe: 'But, ironically, Dr J., when you came back to America, how did the descendants of those same Europeans treat you?' ('Like dirt!' was the reply.) The resemblance to the demagogy of Louis Farrakhan and his Nation of Islam is not coincidental.

Jeffries is one of the 'melanin scholars', together with Hunter Havelin Adams III. He also likes to describe whites as 'ice people' and blacks as 'sun people'. This is the familiar tale of the latitudes once again, but instead of arguing for white mental superiority, it becomes an explanation for white moral inferiority. The basic idea is that Africans maintained a relaxed, generous temperament in their benign continent of origin, whereas Europeans had to develop a mean, calculating character to deal with the demands of the cold north. 'CMWA', the rapper Chuck D of Public Enemy calls it: 'Cave Men With Attitude'.[28]

As well as Diop's *Civilisation or Barbarism*, Jeffries draws upon a book called *The Iceman Inheritance: Prehistoric Sources of Western Man's Racism, Sexism and Aggression*, by an 'at least mostly Caucasoid' Canadian called Michael Bradley. Like Robert Ardrey's *African Genesis* and other popular evolutionary stories of the nuclear era, it proposes that contemporary crises in human society are rooted in the selection pressures of prehistory. Infusing this theme with white guilt, Bradley adapts Carleton Coon's theory of multiregional evolution to argue that the evolution of the Caucasoids differed from that of all other races in one respect: only they passed the *sapiens* threshold in a glacial environment. Their resulting racial character differs from that of the others in one key dimension, that of aggression.[29]

Pulling Freud into the argument, Bradley links Caucasoid aggressiveness to sexual frustration. His vision, like Rushton's, is global but selective. Rushton looks at East Asia and sees a vast sexual backwater; Bradley sees a culture with an epicurean appreciation of polymorphous sexual delights. Rushton's eye is caught by 'undulating rhythms' of African dance, expressing a sexual enthusiasm supposedly peculiar to Negroids; Bradley's by the voluptuous figures of erotic Indian art – Caucasoid bodies animated by Mongoloid culture, he says – and European forms of sexual exaggeration such as corsets and codpieces. He sees the prehistoric figurines known as 'Venuses' as naturalistic representations of a wide-hipped, obese form typical of 'Neanderthal-Caucasoid' women; the physical differences between men and women were greater than in other races, he believes, complicating their relations. Diop, on the other hand, is equally convinced that the 'Venus of Willendorf' statuette displays typically African hair and body form.[30]

As well as demonstrating how easy it is to find examples to fit any sexual stereotype about different human groups, Bradley provides light relief in his self-avowedly racist book by speculating about the *alma*, the Central Asian counterpart of the Yeti. Taking his cue from Boris Porshnev, a Soviet scholar, he suggests that these creatures are Neanderthal relics. So while the Caucasoid race evolved from Neanderthals, the last of the Neanderthals survive in the Caucasus. But Bradley ceases to be comic when he claims that Semitic peoples may be the ' "purest" and oldest Neanderthal-Caucasoids'.[31] The implication is that the Jews are the most aggressive and wicked strain of humanity.

This notion can be slipped effortlessly into the rhymes and cadences of black nationalist declamation, from the podium or over a beat. Khallid Muhammad, Louis Farrakhan's outspoken lieutenant, has worked the trick both on stage and on disc. At Kean College in New Jersey, he improvised on the basic Jewish riff to denounce 'that old no-good Jew, that old impostor Jew, that old hook-nose, bagel-eating, lox-eating, Johnny-come-lately, perpetrating a fraud, just crawled out of the caves and hills of Europe, so-called damn Jew.'[32]

On the introduction to 'Cave Bitch', a track on Ice Cube's album *Lethal Injection*, he turned his attention to women: 'Give me a black goddess sister, I can't resist her. No stringy-haired blonde-haired pale-skinned buttermilk complexion grafted recessive depressive ironing board backside straight up but straight down no frills no thrills Miss six o' clock subject to have the itch mutanoid Caucasoid white cave bitch.'[33]

J. Philippe Rushton would have to acknowledge that the physical stereotype of the 'ironing board backside' is consistent with his own propositions about the greater saliency of secondary sexual character-istics in Negroids, as is the evolutionary 'cave' scenario. The other biological references are exclusive to Afrocentric science, though. 'Mutanoid' and 'recessive' allude to the melanist idea that whites are mutant 'melanin recessives', or albinos. In fact, ordinary whites and blacks both have melanin in their skin, but in differing amounts, whereas albinos cannot synthesise the pigment at all.[34] As it happens, Ice Cube himself distinguishes between albinism in African lineages and Caucasoid cave stock: 'I'd rather fuck an albino. At least I know she's coming from the Nubian and not the Ku Klux Klan that you be in.'

'Yeah, Black Woman, if that's not loving you, I don't know what loving you is,' commented the critic Greg Tate.[35] Like music writers in general, he is less reluctant to attack hardcore rap than most academics seem to be. African-Americans, like most groups under pressure, are reluctant to criticise each other in public. African-American scholars are concerned not to accentuate the distance between themselves and African-Americans of less elevated social status, particularly young ones.

They may be less inhibited when it comes to their fellow academics. Henry Louis Gates Jr, chairman of the Afro-American Studies Department at Harvard University, attacked anti-semitic 'apostles of hate' in a *New York Times* article entitled 'Black Demagogues and Pseudo-Scholars'. He pointed out that black anti-semitism is a weapon in 'the bid of one black elite to supplant another': the challengers are black separatists; their opponents are those 'who have sought common cause with others'.[36] Gates noted that a new edition of *The Iceman Inheritance* carried endorsements by two members of the City College of New York Africana Studies Department, and an introduction by John Henrik Clarke, 'the great paterfamilias of the Afrocentric movement'.

The white left has traditionally made common cause with blacks, but nowadays it is reluctant to support the black elite against the separatist challenge. Much of it now seems in thrall to a relativism that has taken a peculiarly absolutist form. In these conditions, the Afrocentric academy is able to elaborate its fantasies – starting from a mirror image of some of the most egregious and anachronistic regions of white race science – and build its unorthodox street constituency, free from the constraints of conventional intellectual debate.

Among white scholars, the appeal of Afrocentrism is not confined to the Left, either. Having devoted themselves to establishing their thesis that modern society depends to an unprecedented extent on intelligence, assigning power and financial reward accordingly, Richard Herrnstein and Charles Murray offer African-Americans the views of Wade Boykin, who enumerates nine dimensions along which blacks differ from 'the prevailing Eurocentric model'. (Everybody speaks the language nowadays.) According to Boykin, blacks see life as 'essentially vitalistic rather than mechanistic, with the conviction that nonmaterial forces influence people's everyday lives'; they regard music, rhythm and dance as 'central to psychological health'; and 'an orientation in which time is treated as passing through a social space rather than a material one'.

This New Age excursion sits oddly within the remorseless Vulcan logic of Herrnstein and Murray's psychometrics. It makes more sense in the extract adapted from *The Bell Curve* for the *New Republic*, to which Herrnstein and Murray attach reflections they kept out of the book.[37] Ethnic groups all have their own value systems, Herrnstein and Murray observe, in which they accord importance to qualities they feel themselves to possess in abundance. The Irish have the gift of the gab, and give high marks to eloquence; the Scotch-Irish settlers in America 'tended to be cantankerous, restless and violent', but they themselves believed these qualities made them good pioneers.

Warming to this line of folk argument, the authors reminisce about a conversation with a Thai who agreed that Americans have capabilities Thais do not, just as an elephant is stronger than a human. But who, the Thai asked, wants to be an elephant? Similarly, they maintain that people at all levels of cognitive ability generally feel good about their IQ scores. 'It is possible', they conclude, 'to look ahead to a world in which the glorious hodgepodge of inequalities of ethnic groups – genetic and environmental, permanent and temporary – can be not only accepted but celebrated.'

In this vision, power and wealth are distributed according to the objective index of g, but each 'clan' is happy in its lot because of its subjective value system. Although the magazine bills this as 'the case for conservative multiculturalism', Herrnstein and Murray reject the multicultural label. They call it 'wise ethnocentrism'.

8

Tribal Law

WHEN THE TERMS of settlement for the 'new South Africa' began to take shape, conservative Afrikaners lighted upon Zulu separatists as their natural allies. Recognising that the balance of forces now rendered white supremacy untenable as an ideological position, they turned to conservative multiculturalism. The Boers were an African tribe like the Zulus; each needed a political and territorial base for the preservation of their identity. Whites no longer built their case around claims of superiority, but merely of difference. This was European Romanticism in the southern hemisphere. Certain of its adherents, like the eccentric English zookeeper and casino-owner John Aspinall, even saw the Zulus as custodians of a noble warrior tradition which had elsewhere succumbed to the banalities of modern culture. The principle of difference has a far wider applicability, however. In a multipolar world, with a couple of hundred members entitled to a vote at the United Nations, ideologies of supremacy are denied validity. It would not be possible for Radovan Karadžić to justify his actions to the 'world community' on the grounds that his Bosnian Serbs are superior to their erstwhile Muslim neighbours. Instead, he and his colleagues must base their argument upon the proposition that peoples with different cultures cannot live together – while ensuring by means of mortars, snipers, expulsions, robbery and rape that this is true in Bosnia.

The political utility of the idea of difference suggests that there might be widespread demand in the contemporary world for scientific theories of ethnocentrism, just as there was a requirement in the colonial era for theories of racial hierarchy. One or two fringe elements have dabbled with sociobiology, notably the French *Nouvelle Droite* and its principal ideologue, Alain de Benoist. The founding axiom of the *Nouvelle Droite* was not difference, however, but inequality; its roots lay in traditional European racism. Its orientation is expressed in the title of its major organisation, the *Groupement de Recherche et d'Études pour la Civilisation Européenne*, or GRECE, the French word for Greece.

For de Benoist, as for various of his predecessors in the tradition, European civilisation is not Judeo–Christian in essence. He detests both Christianity and Marxism as ideologies based on the myth of a lost state of grace that will eventually be restored. History does not operate on moral principles, he contends, and he draws upon sociobiology as a source of knowledge for an unsentimental understanding of how human society works.[1]

The *Nouvelle Droite* enjoyed a vogue in the late 1970s, thanks in part to the patronage of Robert Hersant's *Figaro* newspaper, which gave the movement a public platform in the *Figaro* magazine. Within the race science underground, the journal *Nouvelle École* forged links with *Neue Anthropologie* and the *Mankind Quarterly*, whose founding editor Robert Gayre was a member of the *comité de patronage*, as at various stages were Hans Eysenck and Roger Pearson.[2]

Although the public soon lost interest in the *Nouvelle Droite*, along with its soupy brew of eugenics and sociobiology, de Benoist always saw his strategy as a 'Gramscism of the right'. Its success would arise from its ability to develop organic connections, over a long period, throughout the tissues of intellectual discourse. Since the fall of the Soviet Union, de Benoist has devoted considerable attention to like-minded souls in Russia. Closer to home, he has orchestrated moves by the radical right to construct novel alliances with anchorless French leftist intellectuals, principally through his journal *Krisis*. He has also cultivated the Greens: among the speakers at GRECE's twenty-fifth anniversary celebrations was Teddy Goldsmith, the publisher of the *Ecologist*.[3] The masthead of the *Figaro* magazine has continued to carry the names of the movement's earlier champions in the mainstream press, Louis Pauwels and Patrice de Plunkett.

The newer ideologists of conservative difference have yet to adopt sociobiology in the same explicit fashion as the *Nouvelle Droite*. This does not mean, however, that the sociobiological vision is uninfluential. Its basic principle is that of self-interest extended through kinship, resolving the 'paradox' of altruism by invoking shared genes. The Marxist biologist J. B. S. Haldane once said that he would not lay down his life for his brother, but would be prepared to do so for three brothers or nine cousins.[4] Sociobiologists apply the same calculus, but without the sense of irony. Extending the arithmetic, they deduce that the principle of 'inclusive fitness' might induce a soldier, say, to sacrifice himself for several hundred or thousand fellows whose sum of relatedness reaches a suitable level.[5]

In doing so, they are working on lines generally accepted in contemporary economic and organisational theory. Sociobiology may not actually need to be explicitly incorporated into ideology, since its assumptions are already so widely disseminated. The anthropologist Pierre van den Berghe claims that the only theoretical models in social science that have any predictive value are those which, like rational choice theory, classical economics and game theory, share some of the assumptions of evolutionary biology. Those that are not reductionist and do not assume individual selfishness, he says, have got nowhere. Among his own predictions, dating from before South Africa held its first universal elections in 1994, was that the desire of the African National Congress to 'wish ethnicity away' would cause the country to descend into chaos.[6]

Meanwhile, contemplating the fate of the mid-century ideal of universal brotherhood, in which every man was his brother's keeper, Donna Haraway sardonically remarks that 'last quarter revisions would read more like "everyone is his/her sibling's banker." '[7]

Van den Berghe sees ethnicity as 'the outer layers of an onion of nepotism, with ego at the core', while relatedness can be seen as 'a set of concentric circles defining declining degrees of relationship: nuclear family, extended family, lineage, clan, dialect group, subethnicity, nation'.[8] Folk logic shares sociobiology's premise of nepotism, and it translates readily to political affairs. 'I like my daughters better than my cousins, my cousins better than my neighbours; my neighbours better than strangers, and strangers better than foes,' observed Jean-Marie Le Pen, leader of the far-right French *Front National*.[9]

Sharing the language and the accepted wisdom of the day, sociobiology has a degree of access to general discourse unusual among scientific disciplines. Its persuasive powers may benefit from the current situation, in which the failure to prevent ethnic conflict is taken as proof that such conflict is inevitable. Sociobiology looks through an evolutionary prism at the human tendency to divide people into us and them, from nations to football supporters. One of its most influential accounts has been developed, in both academic and popular arenas, by Irenäus Eibl-Eibesfeldt, a disciple of the ethologist Konrad Lorenz. He drew upon the notion of cultural pseudospeciation, in which groups devise ways of marking themselves off as separate and consider other groups as less than fully human. The origins of this process can be detected, it is suggested, in the aversion an infant displays to a stranger.[10]

What makes humans so special in their aggression, Eibl-Eibesfeldt argues, is culture. It is normal for species to ritualise conflict between their members, but this typically permits disputes to be resolved without fatalities. Culture, however, has made killing the norm in human group violence. This lends itself to the view that the human animal is dangerously adrift from nature, and that culture is inseparable from murder.

Yet it is also possible to interpret group aggression as a continuation of evolutionary processes, in which history is the history of racial struggle, and the species evolves by internal competition. At the theoretical level, the idea of group selection has been driven from favour by analyses based on the principle that selection acts not on groups, but on individuals and genes. The popular success of Richard Dawkins's *The Selfish Gene* testifies to the ascendancy of the reductionist view. Nevertheless, certain sociobiologists retain sympathy for group selection. They include Umberto Melotti, a member of the *Mankind Quarterly* editorial advisory board, and J. Philippe Rushton, whose journey into race science began with a consideration of the altruism problem. (Ironically, the most prominent opponents of the 'ultra-Darwinism' represented by Dawkins are Stephen Jay Gould and his colleague Niles Eldredge.)[11]

Sociobiology is not homogeneous, however, either theoretically or politically. A collection entitled *The Sociobiology of Ethnocentrism* finds room not only for Melotti, but for Robin Dunbar's critical comments on genetic reductionism and Ian Vine's hopeful views on the possibility of reducing ethnocentrism by changing political structures. Dunbar blames Edward O. Wilson for encouraging the 'arrant nonsense' that sociobiology was about to replace most of the social and behavioural sciences. He argues that sociobiology is concerned with the effect of behaviour on birth and death rates, not the role of genes or other agencies in producing the behaviour. The emphasis on genetic determinism, he suggests, arose because sociobiology's pioneers were population geneticists like John Maynard Smith, or invertebrate specialists like Wilson, who 'are apt to regard even advanced mammals simply as scaled-up insects'.

Ian Vine endorses the idea that ethnocentrism may be rooted in the abreaction to strangers that infants begin to display at around the age of six months. But he regards the resulting ethnocentrism as a weak effect, extreme forms only emerging in certain cultural contexts. Like Rushton, Vine notes that humans may be prepared to subordinate reproductive behaviour to the interests of their group. Rather than seeing this as the secret of ethnic success, however, he observes that 'it is this intrinsic willingness to acquiesce in group decisions that lays us, as a species, so open to cultural manipulation for political or religious ends'.

If ethnocentrism has an innate component, then race may be both an illusion and rooted in biology. This perspective is the basis of Pierre van den Berghe's concept of racism, which he sees as a 'case of culture "hijacking" genes which were selected for different ends . . . and making them serve a totally different social agenda.' But, he continues, there is a biological programme underneath the social agenda. Ethnocentrism is an extension of nepotism, and people use cues such as clothing or speech to assess whether others belong to the group to which their nepotism extends. They have to rely on cultural markers because outsiders usually belong to a neighbouring group and are physically indistinguishable from the in-group. But when physical markers are available, which happens ever more frequently in the age of long-distance migration, they are used in preference. Van den Berghe defines racism as 'discriminatory behaviour based on inherited physical appearance'.[12]

While he accepts that the traditional race concept is untenable, he maintains that genetic differences between populations, such as lactose intolerance and sickle cell anaemia, create behavioural differences between them. 'It is unscientific to pretend otherwise,' he says, also remarking that there are behavioural differences between the sexes and different age groups. So races may not be real, but populations are.

As an example of distinctions based on physical appearance, van den Berghe refers to the three ethnic groups of Rwanda and Burundi. The Tutsi are on average taller than the Hutu, who are generally taller than the Twa, one of the peoples categorised as pygmies. Speaking the year before the Hutu assault on the Rwandan Tutsi in 1994, he observed that stature had been used in the past as 'a quick and dirty basis for sweeping genocidal action'.[13]

Writing in *Anthropology Today* in the wake of the massacres, however, Alex de Waal asserted that 'the differences in physical stature between the groups have been wildly exaggerated: it is rarely possible to tell whether an *individual* is a Twa, Hutu or Tutsi from his or her height'. Noting the colonial view of the Tutsi as tall natural aristocrats, of a Nilo-Hamitic stock that contained a significant non-African strain, de Waal argued that the division between the three groups was actually more akin to the Indian caste system, though a degree of mobility between them was possible. 'The truth is that they were three different strata of the same group, differentiated by occupational and political status,' he declared. The real marker of ethnicity was the identity card, introduced by the Belgian colonial administration in 1926. Anybody with ten cows or more was classified as a Tutsi.[14]

This account can be read as evidence that van den Berghe's model is oversimplistic, or perhaps as the other side of the anthropological coin, denying the physical in favour of the cultural. Ironically, it can also be adduced in support of another of van den Berghe's arguments. He describes himself as an anarchist, while holding proudly to the 1960s anti-racist ideal of equal treatment and institutional race-blindness. One of his favourite hobby-horses is the contention that governments and other institutions perpetuate racism and ethnocentrism when they officially endorse ethnic classification. In the context of academic life in the United States (he is Professor of Sociology and Anthropology at the University of Washington in Seattle) the question turns mostly around the issue of affirmative action. The Rwandan atrocities show that far more may be at stake.

Van den Berghe's ideas were not especially well received when he presented them at the 1993 conference of the Association for the Study of Ethnicity and Nationalism, held at the London School of Economics. Other participants objected to a theory which made racism appear natural, dismissing it as both untenable and dangerous. His insistence on the need to understand the co-evolution of genes and culture appeared to fall on deaf ears.[15]

This was hardly an unprecedented state of affairs. Van den Berghe rejoices in quixoticism, simultaneously challenging the environmentalist dogma of the social sciences and the premises underlying policies directed against institutional racism. In a book published in the late 1970s, he wrote that his universalist ideology had not been altered by his exploration of sociobiology, 'even though I now strongly suspect that my preferences do go against the grain of human and indeed animal nature'.[16]

Vincent Sarich would retort that you can't have it both ways. In the opening lecture of the introductory anthropology course he taught at Berkeley, he recalled what Katharine Hepburn said to Humphrey Bogart in *The African Queen*. Hepburn, playing a missionary, pours Bogart's gin into the river, and a discussion about Bogart's vulnerability to temptation ensues. 'But Missy,' he protests, 'it's just human nature.' 'Nature', replies Hepburn, 'is what we are put in this world to rise above.'

Unlike the liberal sociobiologists, Sarich insists that rising above nature is impossible. He promotes his pessimism aggressively: the transcribed lecture notes which formed his course-reader tell the tale of the opprobrium of his colleagues and the disruption of the class by protestors.[17] His discourses ranged widely, from the molecular anthropology wherein his scholarly distinction lies, to the moral responsibilities of the mentally ill. His outlook has much in common with the worldview of the stereotypical London taxi-driver; in the

United States, a more accurate comparison would probably be with the new generation of populist conservative radio broadcasters.

'It is not that the evolutionary process is capitalistic, but rather capitalism is the first explicit recognition in economics of the evolutionary perspective,' he told his students. For further reading, he recommended the works of the neoliberal economist Friedrich von Hayek and of Garrett Hardin, a biologist. The latter's writings include the observation that 'it is difficult, on rational grounds, to object to the sterilisation of the feeble-minded'. Hardin has been a board member of two organisations, financed by the Pioneer Fund, which argue that the immigration of non-Europeans should be restricted for eugenic reasons.[18]

Faith in the invisible hand often seems to render invisible some of its effects. 'When Professor Sarich was growing up he would not have believed it possible to have young healthy individuals begging in the streets,' the transcription service recorded. 'The idea that individuals could become that irresponsible were [sic] beyond him.' If one shares Sarich's belief that 'what you really need is a market economy and that will solve everything', the correlation of economic liberalisation and the appearance of beggars throughout the streets of the West cannot possibly be causal. But if genes can affect moral fibre, why not macroeconomics?

His views on the allied phenomenon of public insanity were equally forthright. Mental illness was 'at least to some extent a failure of will'. Sarich declared human rights to depend on responsibilities, and illustrated the implications for the insane by reference to Charles Manson, whom he considered not to be human. 'He may have the form of a human being but most certainly not the qualities of a human being. By the same token, many people with the schizophrenic genotype are effectively not quite as human as others with it or without it.'

Sarich also gave the class the benefit of his views on a range of subjects from sex differences to the irrationality of worrying about food additives. Percy Hintzen, a professor of Afro-American studies, was quoted in *Science* as saying that his course on race and ideology 'acted as a release mechanism for the emotional and psychological devastation of having to listen to Sarich'.[19]

In his polemic *The Culture of Complaint*, Robert Hughes remarks scathingly upon the fear of asking students to 'read too much or think too closely, which might cause their fragile personalities to implode on contact with college-level demands'. Hughes argues that, lacking the necessary intellectual skills, students fell back on their feelings. 'When feelings and attitudes are the main referents of argument, to attack any position is automatically to insult its holder, or even to assail his or her perceived "rights"; every *argumentum* becomes *ad hominem*, approaching the condition of harassment, if not quite rape.'[20] In this perspective, education becomes yet another form of therapy. Not so long ago, people of colour in particular were supposed to resist oppression, rather than retreat into the posture of victimhood.

Nevertheless, it is easy to see that after a few weeks of Sarich sandbagging, a student might feel intellectually dispirited as well as morally outraged. It is also conceivable that Sarich's style might alienate students from the discipline as a whole. In terms of public relations, Vincent Sarich is science's worst nightmare. What its critics suspect is the true face of science can be seen in the lecture notes: aggressive, egotistical and unwilling to entertain other points of view. Whatever the strengths of his own hypotheses, he cannot be said to impart the spirit of open-minded inquiry.

On the other hand, lecture notes do not tell the whole story about a teacher or a scholar. Sarich opened one lecture by reading out a letter from a former student who identified herself as a 'feminist woman of color'. She testified that he respected, indeed welcomed, challenges to his viewpoints. In her case at least, the strategy of provocation seems to have proved intellectually stimulating. However, another Berkeley anthropologist, Laura Nader, describes the course as 'indoctrination'. Some of Sarich's colleagues also charged that he was simply not teaching enough anthropology to prepare his students for subsequent courses.

Inevitably, Sarich's evolutionary discourse collided with the theme of race. He dismissed the subtleties of clinal theory with the plain man's assertion that races are recognisable from external characteristics, therefore they exist. The populist streak in Sarich's science is strong, running counter to the insight that a vital part of science is the ability to avoid jumping to conclusions on the basis of superficial appearances. And on the issue of race, there is an illuminating contradiction in the arguments of the 1990 lecture course.

Sarich's most persuasive evidence on the question of racial differences in educational performance is a series of 'Doonesbury' cartoon strips (one of which was quoted earlier) lampooning indolent European Americans who attribute Asian American scholastic success to genes rather than effort. Education was all, or nearly all, he affirmed. Elsewhere, however, he led his students down the path of the relationship between brain size and intelligence. Then he stopped short of spelling out the racial and sexual implications. Even iconoclasts draw the line somewhere.

The reason is probably not lack of audacity, however, but the great fault-line in the hereditarian-neoliberal worldview. As such, it has relevance far beyond the opinions of one maverick professor. The neoliberal has an optimistic faith in self-improvement through effort; the hereditarian has a pessimistic sense of innate limits. As time goes by, and as so often happens when maturity envelops the ideals of youth, the optimist tends to yield to the pessimist.

The 1990 lectures seemed to indicate that the balance of optimism and pessimism was at a transitional point in Sarich's thought. On the one hand, he was affirming the old-fashioned virtues of education. On the other, he had recently reversed his longstanding views on the origins of racial differences. Sarich's pre-eminent contribution to anthropology was made in 1967, when he and Allan Wilson used molecular evidence to show that the lineage leading to the great apes diverged from that leading to humans about seven million years ago. This overturned the wisdom of the day, which held the divergence to have occurred between fifteen and thirty million years ago. The implication that we and the apes are much closer than previously realised has had a profound impact, both in science and in the wider culture.

Once again, Sarich has made what he regards as a crucial deduction about the chronology of human evolution. For a long time, he supported the multiregionalist theory of modern human origins, which considers that racial traits date back to long before the appearance of human beings in their modern form. Although early humans around the Old World differed in form, the remains of their material cultures are indistinguishable. Sarich inferred that racial traits bore no significant relationship to mental ones.

At the end of the 1980s, however, he began to look at skulls, a line of research which has taken him to the natural history museums of London (where he examined fossil hominids) and Vienna (where he studied specimens from Tierra del Fuego).[21] Using a database developed by palaeoanthropologist William Howells to perform a regional comparison of skeletal remains, he concluded that differentiation into the races identified today began as the most recent glaciers began to retreat, about 15,000 years ago. It was not only rapid, but extensive: Sarich advances the claim, also made by John R. Baker, that the range of variation in humans is equivalent to that between other species, such as the chimpanzee and the pygmy chimpanzee. Such species typically take about a million years to separate that far. Sarich concludes the racial differences must have resulted from intense selection pressure, and therefore must be significant. He also notes that the period of racial differentiation also saw striking cultural differentiation.

Other scholars have interpreted the same evidence as implying that racial differences are trivial, and that it was a common advance in mental capacity which permitted human groups to develop their cultures to the point where differences became apparent. Sarich, however, has now declared himself convinced both of the reality of races, and of their inequality.[22]

This brings him back to his opening proposition: that we cannot rise above human nature. He dismisses the concluding thoughts of *The Selfish Gene*, in which Richard Dawkins sugars his pill with the declaration that 'we have the power to defy the selfish genes of our birth'. Sarich acidly comments that Dawkins 'spent 214 pages telling us why that cannot be true', which is what many liberal critics also concluded. Now Sarich has introduced race differences into this vision.

At the outset of his career, Vincent Sarich was a protegé of Sherwood Washburn, whose vision of Man the Hunter created a Universal Man for the post-war era, but their paths long since diverged. In recent years, Sarich's progress has taken him back, beyond the Universal Man, to before the watershed of half a century ago.

PART TWO

9

Socially Unadaptable

IN SEPTEMBER 1993, according to the CTK news agency, the Prime Minister of Slovakia spoke like a man possessed. During an impromptu address to members of his party, the Movement for Democratic Slovakia, in the Eastern Slovak town of Spišská Nová Ves, Vladimír Mečiar was said to have declared that it was necessary to curtail the 'extended reproduction of the socially unadaptable and mentally backward population'.[1]

The shade that seemed to speak through him, across a distance of fifty-odd years and a few hundred kilometres, was that of Dr Robert Ritter, head of the Racial Hygiene and Population Biology Research Unit of the German Ministry of Health under the National Socialist regime. In 1940, Ritter reported that 'the results of our investigations have allowed us to characterise the Gypsies as being a people of entirely primitive ethnological origins, whose mental backwardness makes them incapable of real social adaptation'.[2] The phrases retain their function as racial code perfectly; for any Czech or Slovak listener, the ethnic group in question would immediately be identified by the traits of mental backwardness and social unadaptability.

Unfortunately for Mečiar, the code was internationally transparent as well. In Vienna, the International Helsinki Federation for Human Rights and the Nazi-hunter Simon Wiesenthal issued protests. Wiesenthal suggested that Slovakia's membership of the Council of Europe should be suspended; the Council's Secretary-

General, Catherine de la Lumière, agreed that such action might have to follow if the quotes were verified.[3]

They were not, or not quite. Mečiar vehemently insisted that he had been misquoted, and filed a libel suit against the reporter. A verbatim transcript of the speech was released in his support.[4] Yet for many of his critics, it only served to confirm the offence. It revealed that Mečiar had begun his reflections on the Romani question by referring to the fact that the birthrate of Roma is higher than that of 'whites'.

'So the prospect is that this ratio will be changing to the benefit of Romanies,' he observed. 'That is why if we did not deal with them now, then they would deal with us in time . . .' A matter of us or them. What 'deal' might mean in this context was left to hang in the air, loose talk often being the safest for the speaker and the most threatening for its target.

'Another thing we ought to take into consideration is an extended reproduction of the socially unadaptable population,' Mečiar went on to say. 'Already children are giving birth to children, or grandmothers are giving birth to children – poorly adaptable mentally, badly adaptable socially, with serious health problems, who are simply a great burden on this society.'

What, then, should be done? Robert Ritter had proposed a clear and decisive plan of action: 'The Gypsy question can only be solved when the main body of asocial and worthless Gypsy individuals of mixed blood is collected together in large labour camps and kept working there, and when the further breeding of this population of mixed blood is permanently stopped.'[5]

In due course, the first part of the plan was implemented. But whereas the Nazis' early policy suggestions had envisaged the use of mass sterilisation to fulfil the second part, in practice they moved to a final solution. The *Porrajmos*, or Gypsy Holocaust, remains largely unknown, and the struggle to win it historical legitimacy has led to grotesque arguments over numbers. These are hard to assess, but it has been suggested that a million to a million and a half Gypsies, or between half and three quarters of the entire Romani population, died as a result of Nazi genocidal policies.[6]

Calls for 'death to Gypsies' are now common once again, daubed on walls across East Central Europe. Occasionally, the code words are used. In 1988, a newspaper in the Czech town of Chomutov published a statement attributed to a group called 'Death Squad', which threatened to start 'liquidating all unadaptable Roma'.[7]

Vladimír Mečiar is no Nazi, however, nor even *fascisant*. He is, rather, a populist who has made effective use of nationalist sentiment, in this instance directing it against a group of fellow-nationals. His solution was simply to cut welfare allowances for children.

Western politicians might deplore the fact that a particular ethnic group was identified as the target, but the measure itself is in the spirit of social policy that currently prevails around the world. In the West, the identified targets are individuals who are said to consume disproportionate amounts of public money, particularly lone parents, in polite speech, or 'welfare queens', in the demonology of American conservatism. But those individuals also serve as a means by which the prohibition on attacking entire groups can be circumvented. They stand for an entire social layer, commonly known as the 'underclass'. And, especially in America, 'underclass' is code for 'black'. The better-off majority, known as 'white', resents the spending of its money on a 'black' minority, which lives in appalling social conditions and is feared as a source of crime. In the more antagonistic quarters of white opinion, prejudice hardens into the view that blacks sponge off whites on the one hand, and steal from them on the other. As in America, so in Slovakia, and so on around the developed world.

If the shape of the social problem was essentially familiar, and the proposed solution positively mundane, the question of Mečiar's actual offence remains pertinent. Just what was it about his remarks that put Slovakia in jeopardy, however briefly, of symbolic exclusion from the Europe to which small ex-communist nations so desperately aspire to belong? The offence was threefold. Mečiar's first transgression was to identify the problematic section of society as an ethnic group. His second lay in the use of the phrase 'socially unadaptable', which implied that the Roma were not actually part of society at all. His third was to hint at a biological component to the Romani question.

Although the transcript shows that he did not use the phrase 'mentally backward', as originally reported, that is exactly how the words 'poorly adaptable mentally' would have been understood. One of the most striking indicators of the plight of the Romani population is the rate at which Romani children are assigned to 'special' schools. Although Roma were estimated to comprise 4 per cent of the Czechoslovak population in the 1980s, 27.6 per cent of Romani children were on special school rolls in 1985.[8] The handicap suffered by a significant proportion of these pupils is linguistic: Romani children generally cannot speak Czech or Slovak before they reach school age. Nevertheless, the statistics on mental retardation among them are shocking. A survey of 23,510 children in the Banská Bystrica area, published in 1992, found that while 0.9 per cent of non-Romani children were mentally retarded, the rate among Roma was 21.5 per cent.[9] The authors noted social factors on the one hand, and spoke of the need for 'genetic prevention' on the other.

This is only one example in which scientists have asserted a biological dimension to the Romani issue. Mečiar's remarks show that the scientific discourse can be turned into political rhetoric. And ordinary people's understanding of the 'Gypsy problem' may incorporate scientific notions too. Sandor, a resident of the Romani 'ghetto' in the Eastern Slovak city of Košice, told a reporter from the *Prague Post* that the causes of the extensive mental retardation among many of the children in his neighbourhood were incest and 'degeneration'.[10]

For its part, science is capable of recycling popular notions, as in the title of a paper published by a team at the P.J. Šafárik University in Košice: 'Increased Incidence of Congenital Hypothyroidism in Gypsies in East Slovakia as Compared with White Population'.[11] The folk taxonomy in which 'white' means 'non-Romani' is standard in Slovakia. It takes a difference in skin colour – Roma in Eastern Slovakia are readily distinguished from their neighbours by their darker skin – and elevates it to the status of a racial difference.

The abstract of this paper is strictly scientific in its content and style, yet every statement it makes is vividly political. Beginning in the same style as Prime Minister Mečiar, it makes an observation about the higher breeding rates of the Roma. While Gypsies comprise 9 per cent of the Eastern Slovak population, it states, Gypsy babies amounted to 16.8 per cent of the 90,760 newborns surveyed; and hypothyroidism was nearly three times as common among the Gypsy babies as the rest. It goes on to note that Gypsies have been found to have 'by far the highest coefficient of inbreeding ever reported for any European population or ethnic group', and to suggest that, among Gypsies, congenital hypothyroidism may have genetic causes. Compare this to Mečiar's remarks about the birth of children 'with serious health problems, who are simply a great burden on this society'.

The rhetoric of the paper's opening statements is all the more powerful for its scientific terseness. The scenario proposed is one in which Gypsies – with all the centuries of prejudice that have accumulated around the name – are outbreeding the whites, inbreeding with each other, and producing genetically defective offspring as a result.

It does not speak of a burden to society, but a concern with costs and control is implicit in its source, emerging as it does from that part of the apparatus of the state which applies medico-scientific technologies to the populace. Indeed, its data are derived from a screening programme for congenital hypothyroidism which has been compulsory since 1985. Similar programmes are operated around the world, and their benefits are considered to outweigh their costs significantly. Hypothyroidism in infants causes the condition known as cretinism. Mental and physical development are retarded: the visible marks include a characteristically flattened face, dwarfism and a waddling gait. Unlike the effects of hypothyroidism that develops in adulthood, the damage inflicted by infantile hypothyroidism is irreversible. It has to be detected and treated as early as possible.[12] However, the Košice researchers note that, for some children, treatment could not be sustained 'due to incomprehension of some parents'.

If such obstacles can be overcome, screening saves those affected and their families from a dreadful impairment, and it saves the state a great deal of expensive institutional care. But in Slovakia it is not a purely virtuous system, since screening also serves to provide a statistical foundation for popular prejudice. The programme musters evidence that Romani stock is inferior; the analysis then asserts that this inferiority has a genetic component. What the science is saying, translated into the vernacular, is that the Roma are an inferior race.

Scientific ideas, then, are part of a circuit of exchange linking different sections of Slovak society: the scientific and medical professions, the political sphere, and the popular domain. In this circuit can be seen the faint outlines of a *race science system*; these are only traces and suggestions, it must be stressed, but they are none the less clues to the possible future of racial science.

Race and Slovak science do not interact the way they do because of any peculiarity of the national character. The country's levels of racial prejudice are not grossly out of line with regional standards, though by some measures they may be higher. In a survey conducted across several countries in 1991, an overwhelming 91 per cent of Czechoslovaks said that they disliked Gypsies, compared to 79 per cent of Hungarians, 71 per cent of Bulgarians, and 59 per cent of Germans. A similar poll held in the same year found that 85 per cent of Czechoslovaks said they would prefer not to have Gypsies in their neighbourhood; 76 per cent of Hungarians and 72 per cent of Poles felt the same way. Three years later, the Czech paper *Prognosis* reported a poll indicating that 77 per cent of Czechs had a 'negative view' of Roma, while 30 per cent thought they should be deported or isolated in ghettos.[13]

Both the transnational surveys indicated the Gypsies to be far and away the most unpopular ethnic groups in each country polled. Hostility to the Roma might therefore be considered one of the principal characteristics of racism in East and Central Europe. This hostility is not simply a matter of opinion: attacks on Roma have been common in the region since the collapse of the communist regimes.

The pogrom that took place in 1993 in the Romanian village of Hadareni serves as a ghastly cameo of the forces and beliefs at work. After the fatal stabbing of a Romanian by a Romani neighbour (in a fight between two Romani brothers and seven or eight Romanians), the Romanian villagers took revenge by kicking the brothers to death. They are reported to have been assisted in this by a large number of armed police, who handcuffed the victims first, then watched as the villagers burned thirteen houses, killing a third man.

'We're proud of what we did,' said Maria and Ion, an elderly couple who also stood by and watched. 'On reflection, though, it would have been better if we had burnt more of the people, not just the houses.' 'We did not commit murder,' added Maria. 'How could you call killing Gypsies murder? Gypsies are not really people, you see. They are always killing each other. They are criminals, sub-human, vermin. And they are certainly not wanted here.'[14]

Romanians cannot comprehend why anybody should want Roma anywhere. Since the fall of the Ceauşescu dictatorship, many childless couples have come to Romania looking for babies to adopt, and many of those on offer in the orphanages are Romani. A Mexican-American couple from Texas who specifically asked for a Romani baby, so that the skin shades of child and adoptive parents would more closely match, told the International Romani Union that their request was refused. Why, they had been asked, waste the opportunity of growing up in America on a Gypsy?[15] After all, Gypsies were kept as slaves in Romania until the middle of the nineteenth century.

The *New York Times Magazine* recorded similar views at a state orphanage in Ploieşti. On being told that a foreigner was looking for abandoned babies, Dr Luiza Popescu remarked that most such children came from the 'baby machines', the Roma. 'How could Americans be willing to adopt Gypsies?' she asked. 'The genetics is what matters from the beginning. Ha! Such a child will certainly steal.'[16]

Once again, the observations fall into a pattern. First there is a reference to the high breeding rates of the Roma. In this instance, the phrase 'baby machines' echoes the viewpoint of the Hadareni villagers, who insisted that the Roma were not really people. It is followed by a statement about their antisocial tendencies. This time, there is none of the politician's ambiguity retained in a formulation such as 'socially unadaptable'. The ex-communist Mečiar emerged from a tradition of environmental determinism, which sought to re-engineer Romani behaviour by forcing them to settle down and integrate into the socialist workforce. The phrase 'socially unadapt-able' shows him occupying a grey zone between social and biological determinism, like other underclass theorists around the world. Possibly a change of environment – above all, a reduction in the reproductive rate – might reduce the 'burden to society', possibly not. The tone and the context of his remarks strongly suggest the latter, but he retains a get-out.

Dr Popescu does not need to take such tactical considerations into account. A more categorical statement of genetic determinism would be hard to find, or a more vicious one. Environment is utterly irrelevant, she says; even if the child grows up in the promised land of America, its criminality will be revealed according to the genetic law. Her knowledge of this law may well be shaky, of course, since communist regimes suppressed genetics in the same way as they did the churches, and for similar ideological reasons.

Popescu's declaration is a stark illustration of the power of an idea to endure the passage of history. Joachim Gottfried Herder, the founding father of the Romantic notion of race, believed that each race had its own genius, or *Volksgeist*, which took different forms according to the *Zeitgeist*, or spirit of the age. Dr Popescu adheres to the first part of this principle with her implicit assertion that the Gypsies have an inherent, fundamentally immutable character.

Her phrasing illustrates, by analogy, Herder's idea of the relation between *Volksgeist* and *Zeitgeist*. In the eighteenth century, her claim might have been stated in the Herderian phraseology of *Volk*. In the late nineteenth or early twentieth centuries, she perhaps would have referred to the Italian criminologist Cesare Lombroso's book *L'uomo deliquente*, which discoursed upon the innate criminality of the Gypsies, 'so low morally and so incapable of cultural and intellectual development'.[17] In the late twentieth century, the Romantic tradition of Herder and Lombroso is expressed in the concept of 'genetics'. The spirit remains the same; its form changes according to history.

This is precisely the state of affairs which the public advocates of genetic diversity studies claim has been dispelled by half a century of population concepts. In this usage, 'genetics' preserves the old in the guise of the modern. It combines the strength of a tradition that has spent two centuries permeating the culture with that of a discipline which appears to be nearing its historical apotheosis. Issued by a doctor, it appears to be a scientific pronouncement, with all the additional power that derives from a local history in which challenging authority has generally been unwise, if not impossible. And however often luminaries of the Western scientific elite go on television or issue statements through Unesco to ridicule the idea that race may raise its ugly head in science again, however disreputable or threadbare they might consider such notions, it will all count for very little in the clinics and laboratories where Dr Popescu and her colleagues hold sway.

Outside the institutions, according to Tobe Sonneman of the Romani Jewish Alliance, 'Romanians expressed the pervasive idea that Romanies are genetically inferior'.[18] In the 1980s, the anthropologist Katherine Verdery has observed, many Romanians also expressed the belief that the Securitate secret police were members of another race. They were said to be physically distinguishable from Romanians, but nobody was sure where they came from.[19] An alien race, of mysterious origins, arrives in the land and establishes a reign of terror: it has the flavour of a folk tale, with the Roma in the role of the murdered peasant who returns to wreak vengeance.

Though Dr Popescu's remarks are equally folkloric under the surface, they are an example from outside Slovakia of how professionals with medico-scientific authority can produce statements which may form part of a race science system. In the political domain, Slovakia's neighbour Hungary provides an example to demonstrate that biologistic rhetoric is not a peculiarly Slovak device either. In August 1992, the newspaper *Magyar Forum* published a long article by István Csurka, a writer who emerged as a populist politician in the post-communist era. Entitled 'Setting the Record Straight', it drew upon not just one but both of fascism's favourite theoretical constructs: the international Jewish conspiracy theory, and a biological concept of racial hierarchy. The political scientist Thomas S. Szayna identified it as the first neo-fascist manifesto to emerge in the region after the collapse of the Soviet bloc.[20]

At the time, Csurka was a member of the Hungarian Democratic Forum, then the largest party in the governing coalition. In his article, he claimed that the HDF was the target of an international conspiracy involving the liberal Young Democrat and Free Democratic parties, together with the old communist *nomenklatura*, many of whom were Jewish. As for the state of the Hungarian nation itself, he declared that it was 'unhealthy' to blame the skinheads for everything:

> We can no longer ignore the fact that the deterioration has genetic causes as well. It has to be acknowledged that underprivileged, even cumulatively underprivileged, strata and groups, in which the harsh laws of natural selection no longer function because it would do no good anyway, have been living among us for far too long. Society now has to support the strong fit-for-life families who are prepared for work and achievement.[21]

The notion of 'support' indicates a major difference between Csurka's idea of the useless classes and that of Western underclass theory. Neo-liberalism considers that the 'underclass' is a drain on the achieving classes, who should be allowed to keep and enjoy the fruits of their industry. Csurka, by contrast, is suggesting that these classes themselves need support, despite their strength and fitness. It seems reasonable to infer that by 'society' he means the state. If that is so, then his proposal speaks to a concept of nationhood based on a close relationship between the state and the sections of society that are fittest in the Social Darwinist sense. It bears a superficial resemblance to neo-liberalism, but is in fact authentic neo-fascism.

The use of the formulation 'strata and groups' suggests a conflation, interesting in ideological terms, of class and race. Csurka is obviously using the code for Roma, appealing to the popular notion that they are biologically dysfunctional, but he seems to be identifying them as just one section of the unfit population.

Thomas Szayna comments that Csurka has an 'almost biological understanding of the Hungarian "nation" as a living entity'. The roots of this belief lie in the literary tradition with which Csurka has aligned himself. Between the world wars, the *népi írók*, or *völkisch* writers, espoused a Romanticism which located the Hungarian soul within the peasantry and their traditional culture. From this group emerged the 'Third Road' tendency – anti-communist, avowedly anti-capitalist, and undoubtedly anti-semitic. When Csurka founded his own patriotic movement, he called it *Magyar Ut*, or Hungarian Road; the name was originally that of a populist journal sympathetic to the *népi írók*, which appeared from 1941 to 1944.[22]

Csurka may be more of a writer than a politician. The tactical aim of his article was to embarrass the leadership of his own party with the accusation that its response to the alleged international conspiracy had been ineffectual. In the end, the leadership won: Csurka was expelled. He went on to build his own party around his social base of embittered old people and skinheads, a course that took him into electoral marginality. The Hungarian Truth and Life Party is unlikely to attain its goal of a 'Hungarian Christian Nation State with folk roots'.[23]

His immediate decline does not mean, however, that he or the tradition he represents can be disregarded. As in the West, a radical rightist does not need to mobilise a very large proportion of the electorate in order to be politically influential. The discontents of capitalism that encourage populism and fascism will be felt more acutely in Eastern Europe than in the West for at least a generation. Democratic traditions have very shallow roots; populist ones are part of the cultural patrimony. And it may prove to be in the area of culture that somebody like Csurka can be most effective, keeping the flame of *völkisch* nationalism alive. Paul Hockenos observes that 'for ethnic nationalists, arguments inevitably boil down to science – to states of nature, organic essences, processes of evolution, and biological laws. The biological logic of 'pure ethnicity' unmasks the racial component in such forms of nationalism that inform racist prejudices and discriminatory policies.' Before the war, he remarks, nationalists searched for a 'Hungarian gene'.

The cases of István Csurka and Dr Popescu illustrate the persistence of race-biological language among the elites – literary, political, medical, professional – of other East European states besides Slovakia. But they do not really illustrate the interplay between science and other domains, since they are essentially pseudo-scientific. Popescu might well find a few Western sociobiologists who would agree that a genetic predisposition toward crime exists and is resistant to environmental modification, but her insistence that Romani children are genetically ordained to steal is a *völkisch* statement about racial essences. Csurka's remarks about natural selection no longer working 'because it would do no good anyway' may be effective rhetorically, but are nonsense scientifically.

This is not to say that Slovakia is the only ex-communist country in which the lines of a race science system might be detected. It is just that a number of factors combine to make it a more obvious candidate as a case study.

One such factor is the existence of a large Romani population, though quite how large it is impossible to say. In 1993, before the breakup of the Czechoslovak state, the Council of Europe suggested a minimum of 730,000 and a maximum of 820,000 throughout the country. Yet in a 1991 census, which permitted people to identify themselves as Gypsies, only 33,000 in the Czech region and 81,000 in Slovakia did so. As one citizen explained, 'At the census I declared myself a Slovak because being a Gypsy is reason enough to have other people hate you.'[24]

At least Czechoslovakia allowed its Roma the option of identifying themselves. In Bulgaria, which the Council of Europe estimated to have a Romani population of similar size, the very existence of the minority was denied. The effects of this policy were both shocking and bizarre. Some of them are revealed in an account by Luba Kalaydjieva and Ivo Kremensky, two scientists involved in a national screening programme for the disease phenylketonuria, or PKU.[25] The condition arises from a recessive gene, which causes a defect in an enzyme, which then renders the body unable to convert one amino acid, phenylalanine, into another, tyrosine. Phenylalanine and its by-products accumulate, causing mental retardation and death in childhood. The remedy is relatively straightforward: a special diet free from phenylalanine. Thus, as with congenital hypothyroidism, a neonatal screening programme can indicate treatment which alleviates great suffering and rescues human potential.

A totalitarian state, however, can place peculiar obstacles in the path of such a project. Initially, babies were also screened for galactosaemia, a genetic enzyme disorder which disrupts the metabolism of the sugar galactose, impairing mental development and causing blindness. As with PKU, a dietary solution is available: affected children require a diet free from milk and other sources of galactose. The researchers soon discovered that a deficiency of the enzyme galactokinase occurred at a very high rate among Roma. But they were unable to publish the data 'because it implied admitting the existence of this minority group to the international community'.

The Bulgarian communist regime was not prepared to admit the existence of any other minorities either. Science was expected to uphold this line, as in the conclusions of a study of Bulgarian skulls: 'It should be pointed out that craniometric characteristics of the country's population do not manifest essential differences in the different regions, which demonstrates the ethnic homogeneity of the Bulgarian people.'[26] This paper appeared in 1984, a year marked in suitably Orwellian fashion by a government decree that all Bulgarians must have Bulgarian names. According to some reports, hundreds were killed in the enforcement of the order.

This *ukase* was an act of religious as well as ethnic repression, aimed at Islam. Most of Bulgaria's Muslims are Turks, and these were the real target. But name changes were initially forced upon the Roma, in the 1970s, according to the widely observed principle that repressive measures are applied to the weakest groups in society first. One response which may be open to such groups is satire: some Roma named themselves after Todor Zhivkov, the Party leader, or other prominent citizens.[27] The Bulgarianisation campaign eventually led to an exodus in 1989 of more than 300,000 ethnic Turks, following in the footsteps of 150,000 who had been expelled in 1950. Although overt repression ended with communism, at least one Bulgarian scientist appears to believe the job has not been finished. Professor Georgui Petkov, of the Stara Zagora paediatric clinic, has suggested that the ethnic credentials of the remaining population should be verified by DNA testing.[28]

In a truly totalitarian spirit, the old regime was not content just to change the names. It sought to eradicate all evidence that the Turks had ever existed, ordering that medical records bearing Muslim names be destroyed. The PKU screening laboratory was one of the few institutions that secretly disobeyed the command, but the changes of names made maintaining a proper registry extremely difficult.

None of the other communist states went to such extremes to deny the existence of the Roma. But only the multiethnic states of the Soviet Union and Yugoslavia recognised them as a 'nationality' on a par with other ethnic groups, rather than as a 'minority'. In Czechoslovakia, strenuous efforts to assimilate the Roma population, by force as well as by inducement, were accompanied by denials that the Roma were an authentic ethnic group. 'It is an utterly mistaken opinion that Gypsies form a nationality or a nation, that they have their own national culture, their own national language,' declared the ideologue Jaroslav Sus. Writing in 1958, he dismissed the Romani tongue as 'discordant, stagnant, without written form and therefore without a future, invented by thieves and criminal elements.'[29]

The same year saw the decisive move in the Czechoslovakian campaign, when, as one recent Slovak scientific publication puts it, 'the state gave the Roms permanent dwelling'. At the same time, it declared 'nomadism' illegal. In the words of Jaroslav Sus, 'it would be incorrect, and in the end reactionary, to act against the progressive decay of Gypsy ethnic unity. The way to achieve assimilation does not exclude force, which would tend to remove whatever differences exist.' Among the measures taken to assist historical progress were the removal of wheels from Romani wagons, the killing of their horses, and prison sentences of six months to three years. Officially, however, the outlook was heroically sunny:

> We will successfully solve the Gypsy question, which the preceding social systems were not able to solve. Labour will directly change the Gypsies in miraculous ways. Labour is quicker, the seasons slower. The Gypsies willingly and quickly become accustomed to conscientious work, they adopt trades, they take part in socialist competition, they become shock workers, members and functionaries of the Czechoslovak Communist Party and the Revolutionary Trade Union movement.[30]

Official Marxist-Leninist ideology promoted not so much environmental determinism as environmental transcendentalism. But the persisting inequality of the Roma must be counted in the long list of its failures. Life expectancy for Czechoslovak Roma born in 1980 has been estimated as thirteen years less than for other Czechs and Slovaks. Three-quarters of Romani children fail to complete the basic nine years of schooling. A third of Romani households live in one room; several hundred communities in Slovakia remain without paved roads, electricity or sewage pipes.[31]

The encounter between Soviet bloc communism and the Roma was one of opposite extremes. In medieval Europe, the radical difference between Gypsies and the settled population was not so much their nomadic way of life, but the freedom it represented. The serf lived in bondage and immobility, tied to a feudal lord and a patch of land. Gypsies represented splashes of anarchy upon God's unchanging order. In the twentieth century, they were just as much of an affront to the new order in which God was dethroned in favour of Man. To the apparatchiks, Gypsies were untidy, anarchic and archaic, requiring modernisation in the same way that horses required replacement by tractors.

The process of modernisation, applied to all citizens of the People's Republics, entailed their integration into bureaucratic and technical systems – of municipalities, factories, schools, health services, armed forces. Scientific techniques were deployed to measure socialist progress. Anthropology, widely regarded as ideologically dubious because of its traditional preoccupation with race, achieved a degree of rehabilitation by gathering evidence that children were growing taller under socialism.

As in so many other areas, however, the Soviet-bloc system was very effective in stopping people from doing things, but far too crude to transform the way they thought. Anthropologists had their hands turned to the service of the workers' states, but were not forced to put themselves through the kind of conceptual overhaul that the race issue had provoked in Anglo-American anthropology. In Poland, for example, physical anthropologists stopped performing genetic research in the 1950s, after accusations of racism from zoologists. But Polish physical anthropology held proudly to its distinctive traditions of mathematical typology until the late 1950s, when a younger faction belatedly aligned itself with the New Physical Anthropology promulgated at Cold Spring Harbor in 1950. The key debate between the old and new schools followed the publication of articles from each camp in the American journal *Current Anthropology*, in 1962.[32] As with living standards and car design, the East was about fifteen years behind.

The communist regimes imposed obvious obstacles to the global circulation of ideas, such as restrictions on foreign travel. But a more important obstacle was, perhaps, language. Younger scientists did not routinely learn English in schools; older ones often spoke German. This older generation, bridging the gap of World War Two, may account in part for the lingering influence of German-speaking anthropology across the region during the Cold War period. In many cases, anthropologists who were able to keep up with German journals would have been continuing to participate in the intellectual tradition that had formed them before the war.

There is also a more profound reason for the durability of German tradition. German Romanticism arose as a response to a crisis of national identity in the eighteenth century, reacting against the threat of French cultural influence and, after the conquest of the German states by Napoleon, of French political domination. The principle that each people should come to know and develop its own *Volksgeist* inspired many national struggles in the nineteenth century. But whereas the German crisis of national identity was resolved with overwhelming decisiveness by the unification of the German states to create an imperial power that dominated the continent, national questions arising in the tract of Europe between Germany and Russia remained unresolved, and in many cases remain so to this day. East Central Europe is still a Romantic landscape.

To a significant extent, its Romantic character was incorporated into state socialist ideologies. Recognising that their claims to legitimacy depended on their ability to assume the mantles of nationalism, the communist regimes promoted a wooden folkish-ness. Its characteristic manifestations were state folklore shops, trading in hard currency, and platoons of folk dancers who were trooped out on all possible ceremonial occasions: *Volksgeist* with all the *Geist* drained out of it. Nevertheless, these displays acknowledged that the roots of the nation lay in the peasantry, whose roots in turn were in the land. The German Democratic Republic billed itself as 'the first workers' and peasants' state on German soil'. The soil is a key term of the formulation, not just a literary embellishment.

The idea of the soil is implicit in a remarkable passage by a Soviet anthropologist, V.P. Alexeyev, which was published in 1972. Although Soviet genetics was famously destroyed by the pseudo-scientific doctrine of Trofim Lysenko, which offered a theory of inheritance more consonant with Stalinist dogma than Mendel's, Soviet anthropologists do not seem to have come under the same sort of ideological suspicion over race that was brought to bear on some of their fraternal socialist colleagues. Dismissing the squeamishness of Western scientists about linking peoples, languages and race as 'nihilism' and 'bourgeois liberalism', Alexeyev observes that 'the ethnic origin of the eastern Slavs is felt as a profoundly personal question by Soviet historians. Behind the objective information unearthed by historical, archaeological and anthropological research, they see the blood which has been shed in defence of the fatherland, the hardworking life of the peaceful peasants, the glories of Russian culture.'[33] Blood, land, peasants; *Blut und Boden*, blood and soil.

In the vision that prevails in Central and Eastern Europe, then, nationhood depends on the bond between land and people, hitherto based on the peasant way of life. This implies that the Roma are not an authentic people, and thus that their language must be merely a criminal argot, and so forth. The process of denying authenticity can, and is, taken on to the extreme. Dr Popescu says the Roma are a degenerate people; Jaroslav Sus said they are not really a people; the villagers of Hadareni say they are not really people.

The typological sketch that Professor Ivan Bernasovský includes in the résumé of his research into the blood groups of Roma in Eastern Slovakia – 'Indo-Afghanistan and Iran-Afghanistan admixtures are especially characteristic beside the mediteranoid, oriental, sublapoid, veddoidal and other components' – is a token of his Romantic heritage. Bernasovský acknowledges that modern anthropology recognises only the three 'great races', Mongoloid, Negroid and Caucasoid. His research is based on population concepts of genetics.

Yet he still finds some use for a typological characterisation of the people he studies.[34]

Based at the P. J. Šafárik University, Bernasovský has conducted anthropological research in the Romani communities of Eastern Slovakia since the mid-1970s. Though much of his work is ultimately concerned with the origins of the Roma – and will be described as such in the next chapter – he has also published studies with direct medical implications. Surveying the birthweights of newborn babies in Košice and the nearby town of Prešov, he found that those of Romani babies were significantly lower than the others. In 1961, the World Health Organisation recommended that any baby born at term (after the end of the 37th week of pregnancy) weighing less than 2,500g should be categorised as being of low birthweight. Bernasovský and his colleagues interpreted their data as a reflection of an innate ethnic difference between Romani and non-Romani infants. On that basis, they recommended that the low birthweight threshold for the Roma be lowered to 2,250g.[35]

Their proposal has echoes elsewhere in the literature. One paper from Singapore recommends lowering the threshold for Indian babies on the same grounds, while Northwick Park Hospital in London has instituted a 2,300g threshold for babies of Indian subcontinental descent.[36] A number of British studies have found that such babies are lighter than those of European stock; the difference is typically around 300g.[27]

This is suggestive because, as biologists, linguists and historians all agree, the Roma also originated in India. But it is also too neat to be entirely true. Every shade of opinion acknowledges that environmental influences play a part in determining birthweight. These may include whether the mother smokes, how old she is, how many children she has had, and how tall she is, which in turn depends partly on the environmental influences that have acted upon her. In the Northwick Park study, the average weight of babies of European stock was found to be 3,362g, that of those born to Muslims of Indian subcontinental stock was 3,146g, and for babies born to Hindus, the figure was 2,960g. But after taking various environmental factors into account, no significant difference remained between the Muslim and the European babies, while the difference between the Europeans and the Hindus was reduced from 402g to 190g.

A study on Romani babies conducted across the border from Slovakia, in Hungary, came to opposite conclusions from Bernasovský. Examining 10,108 newborn Roma, it found that their physical development was much poorer than that of other Hungarian infants. When the national figures were arranged according to educational attainment, the pattern resembled that of the Romani group, supporting the view that the numbers were lower because of social, hygienic and educational disadvantages.[38]

Conversely, other studies highlight the possibility that innate factors may sometimes override environmental variation. In Israel, a group of babies born to people of North African origins, who tend to be poor by Israeli standards, were found to be larger on average than those of the population as a whole.[39] Surveys published in the early 1980s found that babies born in Warsaw were about 50g heavier than the newborns of Denmark, whereas the death rate among them was about three times higher, and Poland's per capita national product about three times lower.[40] A Dutch study, entitled 'Differential Birthweights and the Clinical Relevance of Birthweight Standard in a Multiethnic Society', found that, after maternal height was allowed for, Negroid babies were still heavier than ethnically Dutch ones.[41]

Bernasovský's birthweight work poses two questions. Why did he consider it unnecessary, unlike other researchers, to try to control for environmental factors; and what would be the medical and social effects of his proposal?

Taking the latter first, Professor Bernasovský points out that lowering the low birthweight threshold would save the medical services a great deal of costly intensive care. If findings on people of Indian subcontinental stock in Britain are any guide, it appears possible that a particular birthweight may have different implications for babies belonging to different ethnic groups. One study, at a West London hospital, found that among babies weighing less than 2,500g, Indians had a lower mortality rate than Europeans. In the northern town of Blackburn, low birthweight was commoner among babies of Indian subcontinental stock, but a smaller proportion of them needed intensive care, just as Bernasovský would no doubt predict.[42] At Northwick Park, the lowering of the special care threshold to 2,300g had no perceptible effect on perinatal mortality.

In many countries, special care is restricted to babies weighing less than 2,000g. But it may be worth noting the comment of the specialist L. Lubchenco that 'it is a mistake to equate these practices with optimum care'. Lubchenco observes that birthweight criteria in general seem to have more to do with the allocation of resources than with medical physiology.[43]

In this instance, too, a lowering of the birthweight threshold would have the effect of diverting resources away from an already underprivileged section of society. More important, by asserting biological limits to Romani development, it would dilute whatever moral commitment existed to improve the conditions under which Roma live. E. Alberman, writing in the *Journal of the Royal Society of Medicine*, argues that there is a 'close and specific relationship' between birthweight and morbidity as well as mortality. The need to increase birthweight, particularly in underprivileged groups, is further underlined by a number of studies which have shown that starting life in the low birthweight range is associated with subsequent educational, behavioural, sensory and motor problems.[44] The question of mental capacity, which has surfaced elsewhere in the birthweight literature, goes to the heart of the 'Gypsy problem' as it is conceived in contemporary Eastern Europe. Lowering the standards would direct the gaze of medical personnel in the opposite direction.

The social framework within which scientists operate may influence their work in ways that are elliptical and occasionally ironic. In the paper proposing a reduction of the low birthweight threshold, published in Czechoslovakia, Bernasovský does not even mention the possibility that the differences are other than innate. He does, however, make a reference to environmental influences in the account of the work published in *Homo*. Here he notes an Indian study in which average weights and sizes of newborn babies were smaller than those of the Slovak Roma; he observes that 'the significantly better socio-economical conditions of Roms (Gypsies) in Czechoslovakia within the last 3 decades plays an important role'.[45] The period in question was, of course, that of the Czechoslovak Socialist Republic. Comparisons with underdeveloped capitalist states presented no political difficulties, but it might have been injudicious to dwell upon persisting inequalities among citizens of the CSSR.

Nevertheless, although considerations such as these may have subtly influenced the thinking of scientists during the communist period, Bernasovský's belief in racial differences between birth-weights has survived the political transition. It does not seem excessive to suggest that he and his colleagues are predisposed to such views by the intellectual traditions of the region, which in turn are part of a broader cultural heritage. Across the region as a whole, scientists with such an outlook will certainly be confirmed in their disposition by the pressures caused by shortage of money and the drive to adopt Western models of cost control upon public services.

Another subject of Bernasovský's research, however, serves as a useful reminder that predisposition is not predetermination. A prominent theme in the Gypsy stereotype is that of sexual precocity. Romani girls are said to reach sexual maturity earlier than those of European stock, as if in a rush to become the 'baby machines' Romani women are understood to be. Bernasovský found that three times more Roma than ethnic Slovaks become mothers between the ages of fourteen and sixteen; the proportion of such 'children giving birth to children', in Mečiar's words, was 6.1 per cent among the Roma and 2.1 per cent among the rest.[46]

The behaviour of one Romani female in twenty might not seem to be of overwhelming significance, but such is the nature of stereotypes. Their power lies in their ability to appear self-evidently true. A Polish commentator, writing in 1973, remarked that it was incomprehensible 'why early marriage among the Gypsies and premarital sexual intercourse should be treated as a crime since such is the old pattern of their lives, based on obviously physiological differences'. He also observes that, regarding robbery and crimes against the person, 'what might be obvious in our culture, is probably quite incomprehensible for the Gypsies clearly differing in physiological features, biological maturity and cultural character'.[47] With defenders like that, the Gypsies hardly need enemies.

As it turned out, Bernasovský did find physiological differences, but they contradicted the stereotype. The average age of menarche was 13.8 years among Romani girls, and 13.1 years among the others. In other words, Romani girls have their first menstrual periods about eight months later, on average, than their non-Romani peers. It is widely accepted that girls who enjoy a good standard of living reach menarche earlier than poor ones.[48] One study, in the Punjabi city of Patiala, showed an average age of menarche among well-off girls of 12.5; poor girls had their first periods, on average, 1.2 years later. A study conducted by Tadeusz Bielicki in Poland – and published in 1986, when the ideologically exhausted Polish regime had long since given up all but token protestations of socialist equality – found that the daughters of professional people reach menarche earlier than those of manual workers.[49]

Ethnic factors may also play a part, though. In the benign environment of Sydney, the average age of menarche among daughters of immigrants from North-West and Central Europe has been found to be 13.1 years, compared to 12.5 years among girls whose parents came from South-East Europe.[50] But in Eastern Slovakia, Ivan Bernasovský believes that the difference he found can be attributed to social, economic and hygienic factors; the process of sexual maturation is not decisively influenced by membership of a particular ethnic group.

He does, however, believe that ethnic groups may vary in other sexual respects. His team also found that levels of gonadotropic hormones among Romani adolescents were slightly higher than those among non-Romani eastern Slovaks. Citing a paper which found that black girls in America had higher levels of follicle-stimulating hormone than white ones, Bernasovský argues that, in the Slovak case, the difference in gonadotropin levels between Roma and non-Roma is, like birthweight, an ethnic one.

Adhering to the familiar pattern of commentaries on the Roma, the paper on menarche and gonadotropins which Bernasovský presented to an international conference began with observations about the Romani birthrate, and in other papers he also emphasises the Romani 'population explosion'. If the message of the text is not clear enough, that of the picture on the English–language booklet summarising his research on Romani blood groups is unmistakable. Some nineteen Roma crowd to fit into the frame, most of them children or babies. One smiling woman hoists up a baby: maternally proud, perhaps, to sympathetic eyes; or flaunting the evidence of her fecundity, in hostile ones. In the centre of the composition is another smiling woman, like a parody of a Black Madonna, her baby boy twisting his mother's teat in his mouth to keep his eye on the camera. The crowding and the brownness of the skins resonates with the conclusion of the booklet, that the Roma are a people from India, a land where babies are light but abundant.

Within the medical profession, a significant layer of opinion seems to have taken the view that the important thing is not moving the birthweight goalposts, but preventing Romani births from taking place at all. Dr Pavlini, head gynaecologist in the eastern Slovak town of Krompachy, spoke to the international human rights group Helsinki Watch of the 'noble aim of regulating the birth rate . . . the highest disease rate is among these children, and also the highest premature birth rate is among these children'.[51]

Helsinki Watch also interviewed a former patient of Dr Pavlini. 'AD', a Romani woman, went into hospital for an abortion. The doctor told her that the operation might make her temporarily infertile; she was given a paper to sign. When she came round, she was horrified to be told by other women present that she had been sterilised. According to AD's account, Dr Pavlini's reaction to her distress was to slap her across the face. 'Be glad, you cunt, that you won't have any more children,' he told her. 'How many Gypsies do you want to bring to this republic? Hitler was a prick because he didn't kill all of you. What, do you want to overwhelm the entire republic?'

This incident was part of a pattern. Ultimately, Helsinki Watch was investigating whether the pattern amounted to a policy of sterilisation regardless of informed consent. The matter has never been thoroughly investigated, and the evidence is inconclusive. It is, however, not disputed that in the 1970s, state agents – social workers and medical professionals – took exceptional measures to prevent the Roma from breeding. Material inducements were a popular method. Sometimes Romani women would be offered cash in return for being sterilised. Sometimes a sort of barter was employed: for example, the state might furnish a Romani flat as its side of the bargain.

In other cases, as in that reported by 'AD', doctors performed sterilisations under cover of other surgical procedures. Jiří Biolek, a paediatrician from Most, in northern Bohemia, told Helsinki Watch that there had been no official sterilisation policy. However, a 'socially weak' woman might sometimes be sterilised without her knowledge after having half a dozen children. In the theatre, a gynaecologist could always find medical grounds for the operation. 'On the one hand, there are human rights,' he admitted. 'But on the other hand, when you see how these Gypsies multiply and you see that it is a population of an inferior quality, and when you look at the huge sums that had to be paid for the care of these children, it's understandable.'

Other human rights abuses perpetrated under communist regimes have been vigorously investigated, by both locals and Westerners. The testimonies gathered by Helsinki Watch are shocking, not least because they are so consistent with the way the Roma are regarded and treated throughout the region. They are also troubling because they badly need journalistic verification. In part, perhaps, the issue has been neglected because, from the perspective of attitudes to the burdensome classes that are shared around the world, the *ad hoc* campaign was indeed understandable.

One testimony was provided anonymously by a Prague gynaecologist. He described a conference held in Bratislava, the capital of Slovakia, in the late 1970s. One of the papers, given by a Slovak scientist, dealt with the question of Roma birthweight. Roma babies were 100g lighter than non-Roma ones, he said, for two reasons: because they belong to a different ethnic group, and because Roma women have lower IQs and take worse care of their children. The professor from Prague asked how the Slovak had tested the women's IQs, and what Slovak controls he had used to establish the difference between the ethnic groups. No tests had been performed, of course; it was simply taken as obvious. Afterwards, a gynaecologist from eastern Slovakia told the Prague scientist that Gypsies in the Czech lands were more civilised than their Slovak counterparts. He added that he routinely cut the fallopian tubes when peforming caesarean sections on Romani women.

Though this happened under a communist regime, it actually demonstrates that a totalitarian state is not necessary for the practice of racial hygiene. It was sufficient to have a consensus among different sections of the professional elite that action should be taken to limit the 'population of an inferior quality', together with limited bureaucratic provisions, and a population too politically weak to offer effective resistance. While the powers of the bureaucracy and the impotence of the citizenry are obviously greater in dictatorships, the transition to democracy does not seem to have made very much difference in the case of the Roma. Two years after the 'velvet revolution', it was reported that officials were still receiving rewards for meeting sterilisation quotas. Women were still being offered cash inducements to be sterilised, but, in a new twist on the idea of positive discrimination, the offers to Romani women were said to be five or ten times higher than the going rate for the rest.[52]

The sterilisations were originally revealed by the dissident movement Charter '77, which called them 'genocide'. Other communist states did not have such well-organised or vocal opponents; similar practices may have been conducted without ever having been exposed. But on the available evidence, Czechoslovakia was the site of a racial eugenics campaign several decades after the defeat of its neighbour and oppressor, Nazi Germany. This is the last and decisive reason for the concentration in this chapter on Slovakia. Not only can a case be made for the existence of a tentative race science system, in the previously defined sense of a racialised circuit of exchange between science and other domains of society, but there is also evidence for the existence of a loose, *ad hoc* racial hygiene programme. It was not, of course, in the same league as that of the Nazis. Velvet genocide, perhaps.

Charter '77's leading light, Václav Havel, went on to become the first president of post-communist Czechoslovakia, and then of the Czech Republic. He thus found himself a liberal figurehead and source of moral authority atop an institutional structure full of officials eager to harry the Romani minority. 'The Gypsy problem is a litmus test of a civil society,' he has observed.[53]

In this comment, he implies that formal commitments to equality mean little if society and its organic structures – which include its professional cultures and political associations – do not express those principles in their everyday activity. But the Gypsy problem can also be considered a litmus test of a constitution. The citizenship laws adopted by the Czech Republic after the split with Slovakia denied citizenship to large numbers of Roma, provoking expressions of concern from the European Union and the United States.[54]

There is an unhappy irony in the spectacle of a state headed by Václav Havel being criticised, even elliptically, for human rights violations. But that irony is obscured by the character of the Gypsy problem, which does not much resemble that of dissidents like Havel under the old regime. In post-communist Czechoslovakia, official statistics blamed Roma for 50 per cent of all robberies, 60 per cent of all thefts, and 20 per cent of crime as a whole, whereas they were assessed as comprising only 4 per cent of the population.[55] Given the persistence after communism of an overgrown and overbearing police force, it would be extraordinary if these statistics were not skewed by prejudice. But it would be naive to deny that a problem exists, whatever its underlying causes.

The Gypsy problem is also that of cultural difference. The Roma hold to radically different values from the people among whom they live, and this often leads to tension. When Romani refugees from Romania arrived in Germany after the overthrow of the Ceauşescu regime, they were housed in blocks of flats. Although the flats had proper facilities, the Roma would emerge to use the pavements as kitchens and the bushes as lavatories. There can be few countries on earth where such behaviour is less advisable. Yet it arose from Romani concepts of purity. These are organised around the fear of pollution by women, especially menstruating women. A man may be defiled by the presence of a woman in a flat upstairs; he may be rendered unable to drink water from a pipe which a woman has crossed. So, unable to reconcile their way of life with a ubiquitous form of modern housing, the Roma behaved in ways which violated German concepts of purity.[56]

In many respects, however, the Roma have responded with a high degree of flexibility to the pressures imposed by dominant groups around them. Most Eastern European Roma have given up travelling, and many have settled in blocks of flats. They have tended to adopt religions that have a following in the areas they occupy – which has been used as an argument in support of the contention that they are not an authentic people.

But the elements that define a people are not necessarily as visible as nomadism or religious observance. For the Roma, these core structures are those of the family, which supports the complex symbolic order and the system of kinship that interact to constitute the identity of the Romani people. The concept of ritual pollution includes the idea that it can result from even social contact with people who are not *vujó*, or 'pure'. That includes all *gadjé*, or non-Romani people. Marriage to *gadjé* is considered undesirable; conversely, unions between close relatives may be welcomed.

Medical science, of course, takes exactly the opposite view. As described above, biomedical science has gathered extensive evidence of just how damaging the practice of consanguineous marriage among the Roma can be. Although the foregoing also shows how racist ideas can influence scientific accounts of the Roma, it is vital to recognise that the data on recessive disease are real. The classrooms of special schools in Eastern European countries may contain many Romani children assigned to them because they speak the dominant language poorly, or because, in the words of Vladimir Olah, a Romani activist and adviser to the Czechoslovak Ministry of Education, 'when a psychologist sees the brown face and black eyes of a Romany, the child is automatically sent into the special schools'.[57] But the diagnosis of congenital hypothyroidism is derived from hormone levels in a drop of blood. The data will be the same whatever the doctor's disposition towards the Roma.

Racist attitudes and assumptions may help to shape the way that such data are represented: the use of the term 'white' to denote non-Roma in the paper from Košice on congenital hypothyroidism, for example. Roma are beginning to call themselves 'black' as a consciously political act. They are starting to use the label as it is used elsewhere, to signify oppression and resistance.

Instead of 'white', the term 'Caucasoid' may be used for non-Roma, as in a paper on PKU screening by Japanese and Czech authors.[58] A similar usage can be found in a British textbook which, asserting 'ethnic' differences in birthweight, differentiates between 'British Asians' (meaning people from subcontinental Indian stocks) and 'Caucasians'.[59] The term is actually being used as a synonym for 'European'. Its effect is the same as that of calling non-Roma white: it implies a 'racial' rather than a merely 'ethnic' difference.

The usage is not confined to biomedical science. Ian Hancock describes Roma as non-Caucasians whose 'genetic descent is ultimately Dravidian'.[60] This derives from the old idea that equates Caucasians with the 'Aryans' said to have conquered India around 1500BC. Archaeologists such as Colin Renfrew are nowadays inclined to regard this as part of the Aryan myth, questioning whether such an invasion ever took place, and whether the Aryans were very different from other peoples who lived in India.[61] The Roma, like all but a handful of people from the Indian subcontinent, are now considered to fall into the Caucasoid division along with all European stocks.

The traditional story does, however, have a defender in Roger Pearson, editor of the *Mankind Quarterly*, who is fervently committed to the vision of Aryan conquest and dominance over lesser races.[62] Belief in the decisive role of conquering Indo-Europeans – Aryans in a more acceptable formulation – is not entirely confined to the margins, either. The Aryan cult of the nineteenth century had both a German and a French wing, extending the conflict between the two nations to the mythological plane. Its persistence in French culture is illustrated by *Les Indo-Européens*, by Jean Haudry, who locates the birthplace of the Indo-Europeans in central Europe rather than in the Caucasus.[63] Critics have seen the book, part of the *Que sais-je?* series of popular primers, as a sign of rightist leanings on the part of the publishers, the Presses Universitaires de France.[64] Closer still to the centre of power, the former French government minister Michel Poniatowski once remarked that 'only the Indo-European race has the scientific, technical and cultural impulse'.[65]

If a division is racial, it is assumed to override all historical or civic considerations. Like other authors, the Japanese and their Czech colleague speak of the 'host population' and the 'host government'.

To adapt a saying about how long newcomers take to be accepted in English villages, they say the first five hundred years are the worst.

Whether you delete 'white' or 'Caucasoid' and substitute 'non-Roma', though, or delete 'gypsy' and substitute 'Roma', or convert the 'g' to 'G', the numbers in the results table stay the same. By and large, this remains true of biomedical studies even when racist influences operate at subtler levels.

To think of expunging such influences entirely may not be meaningful: it is possible to argue that research upon a marginal social group is bound to reflect the imbalance of power between scientist and subject. But there is nothing in the conventions of scientific writing that rules out the possibility of acknowledging the gap. Science is not only socially embedded, but socially adaptable. Scientists are capable of recognising the difficulties inherent in cultural difference, and discussing them in their work.

A good example of this is 'Disease, Lifestyle, and Consanguinity in 58 American Gypsies', a paper by a Boston group headed by James D. Thomas, which appeared in 1987.[66] The picture it paints of 'lifestyle' choices could hardly be outdone by caricature. Almost all the Gypsies smoked, starting from the age of three, and almost all were obese. Their diet was 'extraordinarily high in animal fat', not to mention sugar and salt. Yet the researchers give an explanation of lifestyle choice which reveals that this wholesale rejection of modern health values cannot be dismissed as fecklessness: 'Exercise was not practised; indeed, such an attempt at personal betterment was felt to be selfish as it took time away from the family, the traditional centre of Gypsy culture.'

Unsurprisingly, rates of vascular disease were extremely high, with hypertension being found in 73 per cent of the subjects. The extent to which the problems found were hereditary was difficult to determine because of the uniformity of the Gypsies' lifestyle. The authors describe their paper as a 'first step', and hope that it 'will prompt others to investigate further these poorly understood but fascinating people'.

Here, perhaps, they reveal the present limits of conventional liberal science, which is learning to listen to its subjects, but has yet to develop ways of working which include subjects as partners. 'These people' – a phrase with an unmistakably colonial undertone – obviously have a richer understanding of themselves than does medical science. Yet they also appear to lack vital insights into their health that medical science could provide. To reconcile the interests of both parties, what seems to be needed is a mutual understanding. At the methodological level, science tends to treat its subject material, human or otherwise, as inert. Human subjects may now be accorded due courtesy as citizens and human beings, but as a research population, they are still likely to remain merely a pool from which a database can be extracted. If they can be recognised as actors in the process of developing knowledge, it may become possible to conduct scientific inquiry as an authentically collaborative venture between scientists and subjects.

There are, of coure, any number of obstacles to such a participatory scientific practice. As the authors of the Boston study put it, 'Gypsies have a unique and cohesive society whose culture may have a detrimental effect on their relationship with the medical profession.' During medical procedures, the Gypsies encountered by the Boston researchers insist on participation, but the measured remarks of the report make it clear that an ideal state of harmonious collaboration has not been achieved: 'Hospitalisation is a turbulent time in the life of a Gypsy, requiring that dozens of family members stay at the hospital, with a male elder typically usurping informed consent from the patient. Demands may be made that departmental chiefs of service care for the patient, and the statements of all care-givers are analysed for possible discrepancies.'

Other accounts confirm the antagonism between Gypsies and medical staff. Both groups have exacting purity requirements, but not the same ones. Medical personnel find Gypsies 'confusing, demanding and chaotic'; the Gypsies view the *gadjé* as 'a source of disease and uncleanliness'.[67] Among the many possible sources of discord is the serious ethical question, raised in the Boston study, of the conflict between patriarchal authority and patient's consent. The one factor unifying the two parties is the basic endorsement the Gypsies give to the medical system. They believe it works, but will only work for them if they continually pick and push at it.

On one hand, then, they are suspicious of institutions; on the other, they are keenly aware of the benefits that such systems may afford them. A similar combination of attitudes is described by E. Mair Williams in her account of PKU testing among Welsh Gypsies.[68] Two cases of phenylketonuria were found in South Welsh Romanichal Gypsy children who were not known to be related: a study was launched to see if this was just a coincidence, or whether the children were actually related through consanguineous unions like those which had been documented among the Gypsies of North Wales. Of 99 matings recorded, 43 were shown to be consanguineous.

The degree of consanguinity is expressed mathematically by a figure called the coefficient of inbreeding, which may be derived in a number of ways. Williams and her colleague Peter Harper calculated it to be 0.017, a very high result which compares on the one hand with values of 0.0063–0.0072, from a presumably typical English rural community, and on the other with the figure of 0.0216 obtained from the Hutterites, an American religious community who act as a sort of gold standard for consanguinity in the scientific literature. It is still well below the European record of 0.05, obtained for Slovak Roma by averaging two figures; 0.017, derived from genealogical analysis, and a massive 0.084, calculated by aggregating identical surnames.[69]

High levels of consanguinity were predictably accompanied by high levels of inherited disorders. The incidence of PKU was found to be about 1 in 40 among Gypsies, compared to 1 in 16,000 among the people of Wales as a whole. Other inherited conditions included 19 cases of non-specific mental abnormality in the offspring of marriages between close relatives.

Williams states that the Gypsies had first to be persuaded that the investigator was not a police officer or public health inspector. Once they had accepted the researchers' credentials, however, they were keenly interested in the findings. 'The family were very concerned about the existence of inherited diseases in their midst and listened attentively to discussions on the subject,' Williams reports. 'All knew the affected children and were very anxious that such a thing should not happen in future generations and therefore, were very conscious

that their children should be taught the disadvantages of consanguineous marriages.'

In the author's account, then, this is a textbook example of
medical genetic research which led to effective health education. It
thus stands in contrast to the Slovak initiative, stymied by what the
researchers tersely describe as 'low co-operation' on the part of the
Roma.[70] Williams has, however, been criticised by Dr Thomas
Acton, Reader in Romani Studies at Greenwich University, for
accepting racist notions about who is and who is not a 'true Gypsy'.[71]
In Britain, one of the most prevalent notions about travelling people
is that they can be divided into Romanies, or true Gypsies, and Irish
travellers, or 'tinkers'. Romantic notions about Gypsy traditions
adhere to the Romanies, while the Irish travellers are regarded rather
as degenerate imposters. Antisocial behaviour by travellers is
attributed to the latter – not least by Romani groups. Williams
reported that her subjects 'very quickly distinguished themselves as
separate from the Irish tinkers whom they considered to be dirty and
dishonest'. She allowed them to define her research population: 'The
decision whether a gypsy was Romany or not was not too difficult to
assess, since the Romany gypsies have very clear-cut ideas regarding
their ethnic identity and keep themselves very much apart from the
itinerant groups of Irish tinkers also travelling in Wales.'

Williams notes that, because of their way of life, travellers may fall
through the net of the national PKU neonatal screening programme.
This applies equally to all travellers, regardless of ethnic affiliation
(the complexities of which will be discussed in the next chapter). So,
by excluding non-Romanies from her study, Williams helps to
perpetuate this medical inequality. It seems possible that, by
identifying with one particular group of travellers, she has become
inclined to regard the 'dirty and dishonest' Irish group as a variety of
the undeserving poor.

This example serves to illustrate the intricacy of the social and
ethnic issues that scientific inquiry needs to comprehend. In
particular, it points to another problem with the tenuous notion of
participatory science: developing a partnership with one group could
lead to the exclusion of another. An issue such as this, however, is

readily tractable. All that is needed, if the critique be accepted, is for the study to be extended to other travellers.

That still leaves the fundamental matter of the data. In this instance, the numbers might not remain the same. It could be argued, for example, that PKU cases among all travellers ought to be aggregated, so as to provide figures that indicate the medical needs of travellers as a whole. But the data for the Romani group itself will be unaffected by conceptual shifts. It still says the same thing as the papers from Eastern Europe: inbreeding among Roma causes a high incidence of mental retardation and other inherited disorders.

Any form of medical science is bound by the ethical imperatives of medicine, and thus has a responsibility to advise upon the medical implications of its findings. In the case of the Roma, this means talk of 'genetic prevention', or eugenics. The advice that follows from the research is that Romani communities should encourage marriage with members of different ethnic groups, while discouraging consanguineous marriages. In other words, the Roma should abandon the values and system of order on which their identity has been based.

The circumstances of some communities may be such that this prospect is not unconscionable. Although consanguineous unions were regarded favourably among the South Wales Romanies, 24 of the 99 marriages documented by Williams included a non-Romani partner, so it was already a familiar phenomenon. What happens to ethnic identity as this proportion rises, remains to be seen. Moreover, the meaning of ethnic identity is affected by socioeconomic and political circumstances. Travellers in Britain certainly suffer discrimination, but on nothing like the scale experienced by Roma in Central and Eastern Europe. In the chaos on the margins of post-communist Europe, the overall chances of creating a productive synthesis between scientific and Romani cultural values look about as good as those of putting Yugoslavia back together again. That is not to say, however, that individual initiatives – such as those undertaken by E. Mair Williams – are necessarily futile.

The inbreeding question is also a problem which looks worse the longer it is contemplated, because after a while it starts to look like an argument about ethnic purity. It is not. There are an infinite number of personal reasons – let alone any other sort – why people should mate across ethnic lines. The history of human races is the history of race mixing, including the intricate subplot of Gypsy kinships in Western Europe. But the attraction between individuals or the interests of families are one thing. An intervention by authority to shape the reproductive patterns of an ethnic group is quite another. Even if the intervention only takes the form of advice, even if that advice is given with the well-being of the community uppermost in mind, even in the unlikely event that the scientist is not funded by the state, that intervention has to be recognised for what it is: eugenics, and racial eugenics at that, to use the old vocabulary.

There is no pat formula to reconcile the medical imperative to help improve health with the truth that, however successfully it struggles against racism, science will find itself at certain points compelled to behave eugenically and racially. It could be said that how science addresses this problem is a litmus test of its role in civil society.

10

The Birth of a Nation

At the same time that science poses an implicit challenge to Romani identity, by detailing the consequences of inbreeding, it makes an explicit offer to help develop a new concept of who the Roma are. In doing so, it asserts a new role for biology in culture.

The field is called ethnogenesis: the formation of ethnic groups. At this point in cultural history, for all the reasons that have been discussed, biological disciplines no longer have a guaranteed place in the study of the birth of nations. Biology is allocated the 'great races', with due caveats regarding the limited value of the concept, but is met with suspicion, or at least scepticism, when it claims to have something to say about ethnic groups. There it threatens to bring the tainted concept of race to a domain now explained in almost exclusively cultural terms. It is trespassing on the domain of historians, linguists, ethnographers and archaeologists.

The primary discipline is history, but its ability to elucidate the origins of the Roma is limited. *The Times Guide to the Peoples of Europe* points out the difficulty: 'For something like half of Gypsy history, obscurity reigns. With few written records to turn to, language and genes must fill the gap.'[1] The basis of the scientific pitch, as articulated by Ivan Bernasovský, is that although linguistics has shown where the Roma came from, it cannot say who exactly their ancestors were. Bernasovský argues that this question may be answered by the study of how genes are distributed among living peoples.[2]

The question of who the Roma are is bound up with the question of what to call them, which also remains unresolved. Most names bestowed on them by indigenous Europeans reflect early misconceptions about where they came from. The European imagination ranged freely: among the places suggested were Atlantis and the Moon.

The Germans and Danish call them *Zigeuner*, the Norwegian version is *sigøyner*, the Polish is *cygany*, the Hungarian *cigány*, and so on. All these variants derive from the Greek *Athingani*, which referred to a Persian sect of magicians and fortune-tellers who came to Greece in the eighth century. In Greek, however, the word is *gyphtos*. Like the Spanish *gitano* and the English 'gypsy' – 'Egipcion', in Elizabethan times – it reflects the other great misapprehension, that the Roma came from Egypt. When the first of the migrants began to arrive in Britain in the fifteenth century, they said that they came from Egypt, or 'Little Egypt'. Their leaders, picturesquely, were 'Lords of Little Egypt'. Elsewhere in Europe, their fellows told the Pope and other rulers that they had come from the Middle East, known in German as 'Klein Egypten'. The Hungarians know them to this day as *Fárao népe*, Pharoah's people. They were also taken to be Turks, Bohemians or Tatars; these labels survive, as in the Swedish *Tattare*.[3]

The trouble with these terms is not so much their inaccuracy, which in any case is now largely buried in etymological history, but the connotations which they have accumulated over half a millennium of mutual antagonism. In former Yugoslavia, depicted before the fall by William G. Lockwood as a unique multicultural island of toleration for Roma, the term *Cigan* was banned as derogatory.[4] Ironically, the outside world has subsequently failed to recognise tens of thousands of Roma as an ethnic group that has suffered from Serbian aggression: in effect, they have no name at all.

As for 'gypsy', 'the word is probably here to stay', observes Ian Hancock of the International Romani Federation. 'It isn't one that is liked very much by a population whose own name for itself is *Roma*.' However, the issue is rather more complicated than that – as Dr Hancock illustrates by describing himself as 'a Romani-speaking Gypsy'.[5] Donald Kenrick, involved at one stage with the London-based Association of Gypsy Organisations, has suggested dividing the European Roma into Kale, who live in Finland, Wales and Spain; Sinte-Manouche, encompassing the Sinte and Manouche travellers of France and Germany, and probably the English Romanichels; and Roma for the rest.[6] In Western Europe, 'Roma' is widely taken to refer to Eastern European Gypsies, who conversely tend to regard all Western European Gypsies as 'Sinte', which leads to the use of the somewhat unwieldy composite 'Roma and Sinti' as an umbrella term. Meanwhile in Romania, the state has decreed that the official name of the Roma is to revert to *Tigan*: untouchable once again. The reason given is the need to avoid confusion with Romanians, though the peoples themselves appear to have no doubts about who is who.[7]

If what Romani groups actually call themselves is taken as the guiding principle, however, the possibility of a simple classification evaporates entirely. Thomas Acton notes that the North Welsh Gypsies are Kalo (which means black), while the South Welsh group whose genes were studied by E. Mair Williams are Romanichal; elsewhere, he observes that 'All the English and Welsh Gypsies that I have ever met would, I think, identify themselves as "Romanies" '.[8] Like anybody else, Gypsies are apt to define themselves in different ways in different contexts.

In recent years, moves have been made to extend the term 'Roma' to embrace all the world's Romani people, perhaps eight to ten million in number. This is a familiar type of contemporary political initiative, seeking to assert a common cause and a common identity, and rejecting labels imposed by outsiders. The Eskimos, understandably dissatisfied with being known to the world by the Algonkian for 'he eats it raw', adopted the name 'Inuit' at the Inuit Circumpolar Conference in 1977. As with 'Roma', it hitherto applied only to part of the total population, the people of Thule.

The particular circumstances of the Eskimo communities made a change of name by conference decision a meaningful one. Similarly, the change of name from 'black' to 'African-American', effected more or less by a collective decree on the part of the black American elite, carried credibility despite the extensive persistence of 'black' in vernacular usage. The status of 'Roma', given the size, global dispersion and heterogeneity of the Romani people, seems less clear. If one term is to be used, however, it seems by far the best, and its new usage is obviously suited to contemporary sensibilities. There is one exception, arising from the historic shift to electronic media, which is moderately comical, but not trivial. Search a database on the term 'Roma', and you will be deluged with every paper published in the capital of Italy. For historians, linguists and geneticists, it may be simpler just to retain the name 'Gypsy'. Back in the real world, the drawback at the anthropological level is that some groups may still consider 'Roma' excludes them.

In asserting ethnic unity, moreover, it glosses over the diversity of the Romani mosaic. Writing of the British Gypsies in the mid-1970s, Thomas Acton remarked that 'they are a most disunited and ill-defined people, possessing a continuity, rather than a community, of culture'.[9] Minimising difference may work when a nation is built within territorial boundaries, but is unlikely to be so effective among a diaspora. The dispersion of Romani communities accentuates the importance of the central question of identity: what is it that the people have in common?

The question can be approached by considering the side-effects of the word 'gypsy'. In the English-speaking world, 'gypsy' has come to refer to a nomadic way of life rather than to an ethnic group. Britain's Caravan Sites Act, passed in 1968, defines 'gipsies' as 'persons of a nomadic habit of life, whatever their race or origin'. In the British Isles, the travelling population comprises both Romanies and so-called Irish Travellers or Tinkers. Unlike the Eastern European Roma, neither group is visibly distinguishable from the general population, nor indeed from the other. The obvious distinction therefore rests in the way of life, and by mutual agreement the term 'Traveller' is favoured over 'Gypsy'.

The effects of this usage have not been uniformly to the Travellers' benefit. A number of pub landlords spotted the possibility of using it as a loophole in the race relations laws. Signs saying 'No Blacks' would have been plainly illegal, and plainly offensive to majority opinion. The signs saying 'No Travellers', a not uncommon sight outside East London pubs in the 1980s, did little to disturb public sensibilities. It appeared merely to be discrimination against people who behave in a particular way, like football supporters, who may also be barred from pubs by standing order.

The 'No Travellers' signs were themselves banned by a ruling by the Law Lords in 1988, which declared Gypsies to be an ethnic group after all.[10] But, like those publicans who continue to display the signs, the British in general seem unconvinced. In a land now dotted with highly visible ethnic minorities, looking different, wearing different clothes, speaking different languages, the Travellers looked inauthentic. Romanies were vaguely understood to have an exotic cultural heritage but, having adopted the culture of the motor vehicle and the television, they were felt to be a tawdry, adulterated version of the real thing. And if they were indistinguishable from Irish Travellers, understood to be no more than an especially degraded strain of Irish people, how could they count as an ethnic entity?

In addition, debates about the peripatetic way of life in Britain have for some time centred on the countercultural dropouts known as New Age Travellers. One effect of this has been to underline the definition of the problem as a behavioural one. But a dimmed perception of an ethnic factor still endures, and it is still dominated by the old criterion of authenticity. Its principal trope is to address the behavioural issue by differentiating between 'true Gypsies' and *lumpen* travellers. According to the 'true Gypsies', any nuisance perpetrated by travellers is the fault of 'tinkers' or other debased varieties; the Romanies' Gaujo sympathisers faithfully repeat such claims.

Traditionally, the distinction has been based on notions of ethnic purity, discriminating not only between Romany and Irish, but between 'pure' Romanies and mixed stock. There are posh-rats, or half-bloods; the phrase is a literal translation of the English expression. There are also didecais: the term is often said to mean 'half-breed'. Although Thomas Acton considers that this explanation is implausible, its racial attribution is the point to note here.

Acton's critique was published more than twenty years ago, but its continuing relevance is affirmed by his observations on E. Mair Williams' genetic studies, which appeared in 1986 and 1987. Williams accepted a definition of Romani ethnicity based on hierarchy, as expressed in the claim that the Irish travellers were 'dirty and dishonest'. This is a traditional race-concept, concerned with notions of purity and superiority. It is equally possible to organise genetic studies around a non-hierarchical notion of diversity – indeed, advocates argue that such a notion is the truth that these studies reveal. Perhaps if scientists were to sample blood from both sides of the demarcation line on Leckwith Heath, in Cardiff, that Williams reports, they might find that Romanies and Irish Travellers were more closely related to each other than to the Gaujo Welsh and Irish. Once again, science would be posing a challenge to ethnic identity.

And a good thing too, many would say. The implications of scientific discourse for ethnic identities are often contradictory. In certain lights, they may deny simplistic race-concepts, revealing unperceived and embarrassing kinships. An example which entertained liberals during the era of apartheid in South Africa was the finding, given as testimony to a parliamentary select committee by a University of Cape Town geneticist named E. D. du Toit, that white South Africans had on average more than 7 per cent 'mixed blood', from Khoisan, Negroid and Asiatic sources.[11]

Exposing the myth of traditional ideas of racial purity is, in fact, one of the things that modern population genetics does best, which is just as well, since it continually needs to assert its anti-racist credentials. But it also lends itself to the assertion of difference, though it has been disinclined to advertise the fact, lest its commitment to the globalistic ideology of United Nations be called into question. At the global level, an enterprise like the Human Genome Diversity Project can claim that its data are bound to underscore the relatedness of all peoples. At the level of the ethnic group, however, the selling point is difference. When Ivan Bernasovský proposes to try to establish the origins of the Roma by genetic inquiry, he is offering to determine something which all Roma have in common, and which differentiates them from other peoples.

With the shift from traditional anthropometry to genetic analysis, significance passed from visible markers to invisible, internal ones. In separating itself from the stuff of folk race taxonomy, dismissing skin colour and facial form as essentially meaningless, population genetics clearly aligns itself with anti-racist ideology. At the same time, it argues that one of the greatest flaws of anthropometric measures is that they are prey to the whims of the environment. Skin colour often changes from season to season; cranial form, as Franz Boas showed, may change from one generation to the next. Genetic markers, on the other hand, are argued either to have no adaptive value, or to speak of ancient and profound selection pressures. As such, they are conceived as possessing an authenticity that escapes visible characters. In fact, they have the qualities that the old anthropometrists were searching for, contriving and then discarding a succession of physically and algebraically exotic indices. At last, science feels confident that it can measure the essence of difference. Naturally, it doesn't put it quite that way.

It is symbolically apt that these essences have been detected in blood, the traditional elixir of racial identity. Blood samples are the science's currency: they contain about thirty different 'systems', starting with the well-known ABO system of blood groups, upon which genetic analysis can be based. Nowadays, the same samples can also be used as the source of genetic material, for the direct analysis of DNA sequences.

The results of these studies do not lend themselves to lay understanding. Not only is variation among blood proteins far more difficult to visualise than the slope of a forehead, but the characters vary independently. Maps of the incidence of different blood groups look like weather maps taken from different months of the year, and in no particular order. Statistical formulae have therefore been devised to combine sets of different readings into single numbers representing 'genetic distance'.

However, having distanced itself from popular understandings of ethnicity by the shift from the visible to the invisible, science then re-inserts its arcane markers of difference into a domain that lay people can comprehend. From the outside, the exciting aspect of human population genetics is that, far from merely trespassing on other scholars' turf, its exponents have woven their findings into dialogues with history, linguistics and archaeology. The implications of this intervention, positing a biological dimension to the humanities, have yet to be fully explored.

Typically, the dialogues take the form of a comparison between theories of ethnogenesis based on linguistic or archaeological evidence, and gene frequency data. Probably the fullest synthesis based on these techniques is the encyclopaedic *History and Geography of Human Genes* by Luigi Luca Cavalli-Sforza, Paolo Menozzi and Alberto Piazza.[12] There is no place for the Roma in its 1,059 pages, however: they are excluded from this history on the grounds that, as an isolate, they require separate consideration. Yet in many respects the Roma are a very attractive population for genetic researchers. Above all, there is a linguistic hypothesis to test. It is the kind of widely disseminated notion that makes for a good textbook example, while at the same time the fact that it is still questioned in some quarters provides a strong rationale for further investigation.

This proposition is far from new. According to an anecdote recounted in Heinrich Grellman's *Die Zigeuner*, published in 1783, a Hungarian preacher named István Valyi encountered three students from India at the University of Leiden and realised that their native tongue was very similar to Romani. They told him that in their native land there was a region called Czigania. Having compiled a lexicon of a thousand words of their language, Valyi found that Gypsies back in Hungary understood them.

Malayalam, the language spoken on the Malabar Coast, is quite unrelated to Romani. However, Grellman surmised that the Indians were Brahmins, and the language in question was actually Sanskrit. It is indeed an ancestor of Romani, although the similarities between the two are not nearly as close as implied in the tale.[13] If the Gypsies spoke a language with Indian affinities, and they looked Indian, they probably were Indian in origin.

This was the first of two inferences that stand as landmarks in Romani studies. The second, which has coloured the discourse on Romani identity ever since, turned not on a lexicon but a single word. It noted the similarity between the word *Rom*, which means 'man', and the Kashmiri word *dom*, which refers to the lowest rank in the Hindu social system. This is the group formerly known as the 'untouchables' or 'pariahs', designated as 'scheduled castes' in the studiedly neutral jargon of bureaucracy, renamed *harijans* – children of God – by Mahatma Gandhi. Turning to the language of struggle, they have now renamed themselves Dalits: the Oppressed.

According to a hymn in the Rig-Veda, the four great divisions of

Hindu humanity arose from different parts of the body of the god Prajapati: the Brahmins from the mouth, the Ksatriyas from the arms, the Vaisyas from the thighs, and the Sudras from the feet. These were the *varnas*, in turn subdivided into *jatis*. *Varna* means 'colour'; in English, both *varnas* and *jatis* are called castes, after the Portuguese word for 'race'. They have something of the character of races, in the old sense of ethnic groups, since they have maintained closed breeding systems for some 3000 years.

The caste system is broadly occupational. The Brahmins were the priests; the Ksatriyas were the military and aristocratic layer, like knights in medieval Europe; the Vaisyas were farmers and merchants; and the Sudras were servants and menials. Below all of these were the 'untouchables'. In his booklet on the seroanthropology of the Roma, Ivan Bernasovský summarises their occupations thus: 'The "untouchables" worked as sweepers and dirt-removers or as washers, navvies, and they prepared wood for cremations . . . They were groomers, executioners and knackers, semi-nomadic smiths, musicians, dancers and performed snake-dances. They were also professional thieves especially skilled in cattle robbery.'[14]

This list serves to illustrate why the equation of *dom* and *Rom* has proved so attractive to European scholars. Apart from fortune-telling, it includes all the major sources of income traditionally associated with the Roma: metalworking, music and dance, casual labour, and theft. In short, it fits the Gypsy stereotype. And it implies that the Gypsy condition is a natural one: whatever land they find themselves in, they will gravitate to the bottom of the social order. The European sense of racial superiority thus draws upon the fatalistic Hindu notion of social immobility.

The widespread use of the term 'pariah' in commentary upon the Roma brings the implication of blame closer to the surface. In contemporary usage, the idea of legitimate exclusion from the collectivity implies not that each person must remain in their allotted station, but that certain persons bring pariah status upon themselves by their actions. It appeals to the perception of the Gypsies as antisocial elements. Milena Hübschmannová, of the Charles University in Prague, criticises traditional 'Gypsiologists' for using terms like 'pariah' without understanding what they really mean.

Against the Gypsiological allusions to the wretchedness of the 'untouchable' condition, Hübschmannová counterposes the proposition that the Roma and the *dom* are descended from the Dravidian people who lived in India before the Aryan conquest, and who created the Indus Valley civilisation two to three thousand years before Christ.[15] The Aryans might have been militarily superior, but as nomads, they had nothing to match the great cities of Harappa and Mohenjodaro, with their grid plans and drainage systems. Hübschmannová suggests that the Roma may be able to claim the Indus Valley civilisation as their heritage, and to see their circumstances as reduced rather than eternally humble.

This scenario falters in the face of Colin Renfrew's scepticism about whether any such conquest actually took place, and whether the Indus people were really any more aboriginal than the Aryans.[16] But to suggest that Romani stock is superior to Aryan adds piquancy to the argument. It twists the tail of traditional Gypsiology, since this is not the first time the Aryans have been the point of reference for an argument about the worth of the Roma. For the Nazis, the Aryan question complicated the Gypsy question. Above all, the Nazis perceived the Gypsies and the Jews as the main sources of alien blood in Europe. But if the Gypsies came from India, they might be Aryans, and therefore affiliated to the highest form of mankind.

'The Gypsies have indeed retained some elements from their Nordic home, but they are descended from the lowest classes of the population in that region,' opined H.F.K. Günther, author of *Rassenkunde des deutschen Volkes*. 'In the course of their migration they have absorbed the blood of the surrounding peoples and have thus become an Oriental, western Asiatic racial mixture, with an addition of Indian, Mid-Asiatic and European strains. Their nomadic way of living is a result of this mixture . . .'

In 1936, Robert Ritter's *Rassenhygienische und bevölkerungsbiologische Forschungsstelle*, or Racial Hygiene and Population Biology Research Unit, was established to determine whether Gypsies were Aryans or *Untermenschen*. It was somewhat confused, however, by Heinrich Himmler's notions of what constituted a 'true Gypsy'. The Nazi secret police chief divided Gypsies into tribes: the Lalleri and Sinti should be declared Aryans, he considered, but the Romanies were racially impure because they mixed with other groups before arriving in Germany. In the end, the badges on the camp uniforms simply said 'Z', for *Zigeuner*.[17]

Milena Hübschmannová bases her argument on a conjecture made by Ivan Bernasovský, about the significance of an allele known as Fy. This gene, part of a blood group system called Duffy, is prevalent among Africans, particularly those living in a belt stretching from the Gulf of Guinea to the eastern seaboard. It is also extremely common in southern Arabia; elsewhere in Western Asia, including the western edge of the Indian subcontinent, it occurs at levels that are much lower but still significant. As far as can be ascertained, it appears to be completely absent from peoples settled in Europe since before the Middle Ages. It has not been detected in most of the Romani groups tested either – except for two communities in Slovakia, in which Bernasovský has recorded Fy frequencies of 17.1 and 16.9 per cent.[18]

The Duffy blood antigen, one of the proteins that stud the surfaces of red blood cells, is the gateway to infection for a parasite called *Plasmodium vivax*. This organism attaches itself to the Duffy antigen, and then penetrates the cell, causing a form of malaria. But the Fy variant of the antigen offers a poorer handhold to the organism. *P. vivax* is not found in regions where most people have the Fy variant, suggesting that the allele is effective in denying the parasite a host population. In America, where there seems to be no selection pressure to maintain high Fy levels, the Duffy group acts as a racial marker. The degree to which 'European' Duffy alieles occur in the African-American population is taken as a measure of white admixture.[19]

For Bernasovský, the presence of Fy in some Slovak Roma suggests not so much Indian roots, but roots in a malarial region of the subcontinent. That would exclude the Punjab, traditionally mooted as the point of departure (and formerly accepted as such by Bernasovský).[20] Malaria thrives in agricultural regions and river valleys – such as that of the Indus, perhaps. Bernasovský proposes to investigate the possibility in a research trip to India, where he will try to elucidate just which *jatis* are most closely related to the Roma, by examining blood groups and DNA polymorphisms. This project, he stresses, is not merely of academic interest, but 'may have important and wide-ranging social and political implications'.

He includes Milena Hübschmannová among the colleagues he would like to accompany him, and has drawn upon an argument she makes about an undesirable polarisation between 'gypsiologists' and Romani scholars. On the one hand, he observes, 'gypsiologists' 'indirectly justify the widespread condescending attitudes' towards the Roma by stressing that they were outcasts in India. On the other, Romani scholars understandably 'glorify the ancient history of Romanies, often without solid evidence'. The antagonism among academics exacerbates the antagonism between the ethnic groups they represent, whereas, Bernasovský argues, science should encourage tolerance. Biology, he implies, can offer the objectivity that the humanities cannot.

Contemporary Romani scholars are capable of far more than romantic partisanship, however. They have also produced a more complex historical account than that hitherto favoured by Bernasovský. While he acknowledges that the origins of the Roma may be heterogeneous – a scientist studying genetic variation could hardly fail to be aware of the possibility – his publications have emphasised the 'outcast conception'. His favoured scenario has been that the ancestors of the Roma, being the 'socially weaker part of the population', were driven out of their region of origin in Central India, moved to the Punjab, and were thence forced out of the subcontinent by the Arab invasions of the seventh and eighth centuries. Over the next seven hundred years, they moved westwards in two streams; one passed through Syria into North Africa, the other reached Europe by means of Armenia and the Balkans. Their tracks are left in the dialects of the Romani language, which has adopted words from the regions through which the people passed.

Whatever the original condition of the Roma, then, and whatever their relationship to the builders of Harappa and Mohenjodaro, their Romani identity was based on weakness and unwantedness. As long as Roma are taken to be no more than an offshoot of *dom*, this assumption will stand. Bernasovský's planned research has the potential to modify his historical assumptions, and to enrich the texture of the historical portrait of the Roma. But researchers in the humanities seem to be ahead of him. Ian Hancock gives an account which begins with the contention that, although the groups who became Roma certainly included *dom*, they also included representatives of other walks of Hindu life.[21]

Hancock endorses linguistic and historical arguments against an early date for the exodus. In his scenario, the Roma were conceived in the invasions of north-western India by Ghaznavid forces, Muslims of Afghan and Turkic stock, between 1001 and 1027. To oppose these incursions, the Indian potentates created a new people, the Rajputs, from northern Indian non-Aryan groups, and promoted them to the Ksatriya *varna*. They were serviced by a population drawn from the Sudra *varna* of menial occupations, and 'untouchable' groups with useful skills.

Despite offering fierce resistance, the Rajputs were eventually beaten. Some clans, together with their camp-followers, headed out of the subcontinent. 'As they became more and more remote from their homeland,' Hancock writes, 'their shared Indian identity, we may hypothesize, overcame whatever newly-acquired jati or caste distinctions might have divided them socially; and in time, the population became one.'

The idea that the proto-Roma went out of India in a blaze of glory has an obvious appeal to romantic cultural nationalism – especially as a counter to the insinuation that they were pariahs driven out with their tails between their legs. Hancock says as much himself: 'The knowledge of having descended from a proud warrior caste and having retained so much of its Indian heritage across the centuries, has become vital to Romani identity and the growing spirit of reunification.'

Identification with the land of ancestors, the *Baro Than*, is a double-edged weapon for Roma. In the old days, under communism, Czechoslovak Roma composed songs about Indira Gandhi and going back to India. Now 'Roma go home to India' is written on the walls.[22] In February 1995, a sign bearing such a slogan appeared outside the Austrian town of Oberwart, 100 kilometres south of Vienna, on a road leading to a Romani settlement. When four local

Romani men approached it, they were killed by a booby-trap bomb.[23]

Back in India, the nomadic Banjara people believe themselves to be Roma who never actually left home. One Banjara tribe, the Lobanas, has been identified as Romani on linguistic grounds. The blood groups of the Lobanas have been investigated by Sarabjit S. Mastana and Surinder S. Papiha, who subsequently widened the scope of their research to consider the range of biological diversity among Romani peoples from India to Wales.[24] They conclude that the European Roma divide genetically into two broad groups, the eastern and the western. There are five or six million Easterners, compared to about a million Westerners, and the balance of numbers reflects the finding that the Easterners are the core population, with close affinities to the Lobanas and another Indian nomadic group, the Sikligars. By contrast, the data from England, Wales and Sweden showed that the Gypsy groups were more like the 'native European' populations.

While they bear an overall resemblance to the Indian nomads, however, the Eastern European Roma vary widely among themselves. Like Ivan Bernasovský, whose work also emphasises the details of Romani genetic variation, Mastana and Papiha argue that variation among these groups is best explained by genetic drift and founder effect. In other words, the Roma are an atomised people, divided into small bands which have become genetically isolated from each other as well as from other populations. Rare genes play major roles in such breeding units, with alleles for recessive traits coming together and finding expression through consanguineous unions. Breeding units draw upon the stock of variation embodied in the founders of the group; the smaller the founding population, the less the variation. The pool of diversity is thereafter vulnerable to twists of fate, such as a fire or an outbreak of disease in a particular household.

Among the Western groups, by contrast, Mastana and Papiha believe that the variation is best explained by interbreeding with local populations (though Bernasovský disagrees). At this point, it might be objected that the genes are telling us little we do not already know. After all, they are only underlining what can readily be deduced from the fact that British Gypsies look like almost any other long-established Britons, while Slovak Roma look distinctly like Indians. But it is always important not to take folk race-taxonomies at face value, and to compare them with systematic evidence.

The genes also have something significant to contribute to the idea of isolation. Again, it is probably clear to most people in Eastern Slovakia that many Romani groups live in isolated communities and do not mix with each other. The weakness of observations like these, however, is that it may not be safe to assume from them that the Roma have always lived like that. The genetic data add depth to the insight by confirming that isolation must have persisted over many generations.

By the same token, the power of the genetic studies arises from their interconnections with other forms of knowledge. They may corroborate the findings of other disciplines. The Indian origin of the Roma is an impressive case in point, unanimously affirmed by studies from Spain to Slovakia. Genetic studies may fill in the gaps left by other disciplines, such as the question of who the closest relatives of the Roma are, and were.

They may also deepen the appreciation of difference by showing another aspect of it. The understanding that the Roma are culturally Indian may be enhanced by the understanding that they are biologically Indian too. For Romani scholar-activists like Ian Hancock, frustrated by the widespread refusal to recognise the Roma as an ethnic group, the biological data help consolidate the argument that 'we are an Asian people, speaking an Asian language, supporting Asian cultural values, having an Asian worldview, in a "white" environment'.[25]

It is not simply a matter of two parameters being more impressive than one, but of increasing knowledge by making connections between the subjects of different disciplines. Insight arises, for example, by considering the scientific literature on inbreeding in the light of scholarly demonstrations of the similarities between Romani and Indian culture. Many Gypsy groups have traditionally been associated with particular occupations, and it may be that this is a continuation of the *jati* caste system, in which groups practising a specific trade married among themselves. If so, then this demonstrates that inbreeding cannot be attributed solely to xenophobia and ignorance of the medical consequences. It also suggests new lines of ethnogenetic research. The knowledge that grows out of such interfertilisation is more than the sum of its parts.

There are obstacles, however. Comparing the work of E. Mair Williams and the social anthropologist Judith Okely, Thomas Acton argues that both suffer from the split between social and physical anthropology. 'Williams writes about matings, Okely about marriages,' he observes. 'Williams writes about hereditary diseases, Okely about cleanliness and health taboos. Williams writes about death as mortality statistics; Okely writes about funerals . . . They *both*, however, suppose these concerns can happily be divorced from one another'[26]

Acton identifies the source of the split as racism, or rather the fear of it. Both anthropologists, he suggests, may be scared to make connections between the social and the physical, lest they end up drawing racist conclusions. Ironically, this leads Williams to build her dataset on a concept of the 'true Gypsy' which dates from the age of scientific racism. Okely denies that biological perspectives are relevant to questions of ethnic identity, Acton argues, and so is left without adequate means to explain the relationship between British Travellers and Gypsies elsewhere. Both anthropologists are still encumbered by the legacy of scientific racism.

They would be liberated if the two great currents of anthropology, the social and the physical, were to flow together once again. As Acton points out, this was inherent in the original nineteenth-century vision of a general science of humanity. These lessons do not apply simply to Roma, as the following chapters in this account will show. Scientists affiliated to other oppressed ethnic groups are likewise concluding that human biological diversity needs to be explored, not denied.

II

The Seacoast
of Macedonia

IN DISCUSSING THE ethnic convolutions of Macedonia, the most Balkan region of them all, it is commonplace to observe that a *macédoine* is a finely chopped salad. The list of ingredients is variable, and intensely disputed, but never has it been so long as in 1918, when one of the fronts of the First World War ran through the region. A rainbow alliance was assembled on the *Entente* side of the line, deploying manpower not only from the homelands of the European powers, but also from their colonies in Africa and India. To Ludwik and Hanka Hirszfeld, he of the University of Zurich and she of the Royal Serbian Army's Central Bacteriological Laboratory, this variegated force offered an opportunity to create a new race gallery, based not on the form of the body but on blood.[1]

The Hirszfelds' attempt to 'attack the human race problem' was based on the discovery, made by Karl Landsteiner and his students in Vienna at the turn of the century, of the ABO blood group system. Were there racial variations in the distribution of these groups? The Hirszfelds, who themselves were from Poland, obtained samples representing the English, French, Italians, Germans, Austrians, Serbs, Greeks, Bulgarians, Arabs, Turks, Russians, Jews, Malagasies from Madagascar, Negroes from Senegal, Annamese (Vietnamese) and Indians. The Jews were refugees from the Macedonian town of Monastir, displaced once again after their arrival from Spain about four hundred years before; Macedonian Muslims were used to stand for the Turks, despite the certainty of a 'large admixture of Slavic blood'.

In a paper first unveiled at a reading to the Salonika Medical

Society in June 1918, the Hirszfelds arranged their results according to the respective frequencies of groups A and B. The results, as they noted, were remarkable. In all the European groups, the prevalence of A was above 40 per cent; in all the others, it was lower. Group A gradually diminished going southwards and eastwards, with Arabs, Turks, Russians and Jews forming an intermediate type between the Europeans and the 'Afro-Asiatic' type. Conversely, group B was at its highest outside Europe, diminishing along a north-westerly heading. Its lowest incidence was among the English.

The Hirszfelds were struck by the fact that, elegant as these gradients appeared on a graph, they were at odds with certain aspects of traditional race classification. If blood groups had been aligned with familiar anthropological arrangements, the highest B levels should have been found among the Africans, or at least the Annamese. Yet these were only the third and second highest; they were far outstripped by the prevalence among the Indians, who were considered to be anthropologically closest to Europeans. Moreover, 'the Russians and the Jews, who differ so much from each other in anatomical characteristics, mode of life, occupation, and temperament, have exactly the same proportion of A and B.'

The blood groups thus defied both standard anthropometric measures and the Romantic race concept in which physical characters were inseparable from psychological ones. The Hirszfelds were convinced that blood revealed a deeper level of truth than body dimensions and form, a conviction shared by their descendants in the tradition of genetic diversity studies. They announced the discovery of 'biochemical races', and suggested a double origin of the human species, the B race arising in India and A race in Northern or Central Europe. Their contemporaries were frustratingly slow to accept their ideas. Later, the Nazis dabbled with the use of blood groups for racial classification. In the following world war, the Hirszfelds found themselves not in an ethnic *macédoine*, but in the enforced purity of the Warsaw Ghetto. There, Ludwik Hirszfeld organised a laboratory and gave illegal lectures on medicine, including one on the subject of blood groups and race.[2]

In 1918, the researchers and their subjects had been brought together by the agencies of war, empire, expulsion and flight. It is now possible to detect the traces of such phenomena by using the same sort of techniques, mapping the rise and fall of gene frequencies across populations. Blood now offers many more genes to choose from, including a dozen or so blood groups, a similar number of red cell enzymes, and various proteins. In recent years, techniques for the direct examination of DNA have come into use.

As the capabilities of population genetics grow, its advocates become increasingly outspoken in their claims of explanatory power. 'Our past is a combination of culture and biology,' says Sir Walter Bodmer, a prime mover of the Human Genome Project. 'One without the other is only half the story.'[3] In arriving at a stage where such a claim is credible both as a proposition and the basis of practical research, genetics has intertwined itself with human history and geography, from the grand panorama of continents to the genetic view from the village green.

The Hirszfelds' tradition has been maintained in the Balkan region, whose 'disordered topography', according to Huxley and Haddon in *We Europeans*, 'has much to do with the equally disordered distribution of its mixed population, and helps to account for the bitterness of their animosities'. Pavao Rudan, of the University of Zagreb, has studied human microevolution along the Dalmatian coast, and has also been involved with similar studies in the Djerdap region of the Danube valley in Serbia. In these surveys, blood group data are integrated with other characteristics, such as body measurements, cardio-respiratory measurements, bone dimensions and dermatoglyphs – finger and palm prints. All these are brought into dialogue with history. The Adriatic island of Korčula spent four hundred years, until the end of the eighteenth century, under the control of the Venetian republic; the Pelješac peninsula, just a mile away, belonged to the Dubrovnik republic. This, Rudan and his colleagues suggest, is the source of the biological differences they found between the people of Pelješac and the offshore islands today.[4]

The report of the Djerdap research appeared as part of a series entitled 'Ethnoanthropological Problems', launched in Belgrade in the mid-1980s as ethnoanthropological problems began to loom large over a moribund Yugoslavia.[5] Its narrative theme is that of a microcosm of Serbian history, a 'transitory region' whose population periodically leaves during various crises and later returns. Biologically, the population is found to be homogeneous. One of the reasons offered is the fall in the birth rate, known in this area and in the Vojna Krajina borderlands of Croatia as the 'white plague', resulting from emigration, war and economic crisis. This, it is reasoned, clears a space for immigration, which brings in a flow of genes that helps to homogenise the population.

A text such as this can be read not just as a scientific paper, but as a reflection upon national history and contemporary concerns. In its depiction of genetic variation as a perennial ebb and flow, driven by the hydraulics of war and other crises, it serves as a reminder that, from the mid-1980s, people in Yugoslavia knew what was coming.

Just as history and politics were drawn into scientific discourse, scientific language occasionally cropped up in nationalist rhetoric. Scientists themselves, however, are not usually prominent in the vanguard of nationalism, particularly the sort which has emerged after the collapse of the Soviet system. Since the characteristics which most obviously differentiate neighbouring ethnic groups are cultural rather than biological, linguists and historians are natural leaders in the politicisation of these differences. Thomas Gamkrelidze, for example, has long been a leading authority in the debate over the origins of the Indo-Europeans, which he considers to have been somewhere in the region of his native Georgia. As well as being a professor of philology, he became the chairman of the parliamentary foreign affairs committee early in Georgia's turbulent independence. His scholarly work includes arguments that the Abkhazians, who fought a war of secession from the Tbilisi government, are not an authentic ethnic group with a separate identity from the Georgians.[6]

A scientist, if so minded, may none the less play an active part in a nationalist cause. While a number of prominent figures in the Bosnian Serb leadership emerged from those disciplines ironically known as the humanities, the woman known as 'Vice-President' is a biologist. Biljana Plavšić stands out among her colleagues not only as a scientist, but as a nationalist whose views are extreme even by Bosnian Serb standards. She does, however, have her admirers. Yuri Lochtchich, one of the Greater Serbian movement's numerous

Russian groupies, considers her 'the incarnation of European beauty'.[7]

On at least one occasion, Plavšić has used her scientific credentials to lend authority to her political opinions. 'The Bosnian Serbs . . . have developed a highly-tuned sense which permits them to sense when the nation is in danger and to put a defence mechanism into operation,' she observed to the Serbian newspaper *Borba*. 'In my family, we always said that the Bosnian Serbs were better Serbs than the Serbian Serbs . . . I am a biologist and I know that the species best armed to adapt and survive are those which live close to other species which represent a danger to them.'[8]

By arguing along these social Darwinist lines, she is able to refer to the differences between Bosnian Muslims and Serbs in terms of differences between species. She has also made the explicit claim that Serbs are genetically superior. When a journalist asked Plavšić for her reaction to a comment by Ejup Ganić, the Bosnian foreign minister, who said that the Bosnian Muslims were of Serbian origin, she retorted: 'What can I say? It is true. But what we have here is genetically defective material, that has converted to Islam. And now it becomes denser from generation to generation. It is becoming more pernicious and malignant, it expresses itself ever more strongly, dictating thought and action. All that is already implanted in the genes.'[9]

Conversely, the Serbian genotype preserves mankind's highest instincts, according to Radoslav Unković, director of the Establishment for the Historico-cultural Heritage of the Serbian Republic. Writing in the newspaper of the ultranationalist Serbian Radical Party, he observed that 'even in the gravest and darkest times, we have never lost our cultural tradition and the genetic impulse towards civilisation'.[10]

These examples of biologistic rhetoric are among a number collected by Ivan Čolović, a Belgrade anthropologist who has taken an interest in the florid language of nationalism. It is a rich vein to work. Among the contributions to Serbian folk-genetics is one from a correspondent for the newspaper *Vojska Krajina* (Krajina Army), published in the Croatian border region controlled by Serbian secessionists. When the reporter and his companions arrive at a refugee camp, the children 'spontaneously and unexpectedly' mount a cultural programme for the visitors. The highlight is a song, supposedly of their own composition: '*We have donned the khaki uniform/To defend the village where we were born/Smoković, place where we first saw the light of day/We will not let you fall into the hands of the Ustasha/Oh you fascists, your mother is wretched/You will not forget Captain Dragan.*'[11]

Not only do these children, one of whom is only five years old, speak the language of their elders, but they have instinctively adopted the rhythm of their ancestors. The song is in the decasyllabic meter typical of nineteenth-century Serbian epic verse. Yet the refugee children 'have no idea what a decasyllable is, nor are conscious of having used it when they composed this song, it is innate in them, inscribed in their genetic code'.

It appears that spontaneous demonstrations remain a feature of formerly communist countries, rather as the Blessed Virgin Mary is wont to make regular appearances in Catholic ones. Post-communist propagandists have an unprecedentedly wide repertoire of images and devices on which to draw. The *Vojska Krajina* correspondent has contrived to link the rhetoric of the Second World War anti-fascist struggle to an explicitly Serbian tradition of patriotic culture, one whose continuity is preserved in the genes. There could be no clearer illustration of the way in which the idea of the gene can be used to dress the Romantic concept of racial character in contemporary scientific terms.

An obvious objection to this example is that it may be no more than a rhetorical conceit, which the author does not intend to be taken literally. But the distinction between literal and figurative is probably not a meaningful one in such circumstances. Any nation at war is likely to succumb to a regime of authoritarian hysteria, in which the questioning of any expression of patriotism is regarded as unpatriotic. Even to consider whether such a claim is literal or figurative would count as questioning, and therefore a disloyal act. The result is rhetorical hyperinflation.

There is also an apparently endless supply of belief systems which refer to biological notions and which therefore appear, at least in the eyes of their sympathisers, to mediate between science and other domains of discourse. Another example from Serbia suggests that pseudo-science may be a useful device for obscuring the distinction between biology and culture, and for reifying nationalistic myth as physiological reality. Like Radovan Karadžić, Jovan N. Striković is a psychiatrist and a poet. Drawing upon Jung's idea of archetypes, Striković claims that the 'collective mind' of a nation bears physical, heritable traces of great historical events:

> According to a phylogenetic law, the psychic structure of the personality also preserves the mental imprint of ancestors, in the same manner as its anatomical substrate. When a nation attains a certain level of spiritual development, the personalities which have marked their epoch reappear; it is not a matter solely of a 'metaphysical' substance, but of a concrete and physical substrate present in the central nervous system of the members of a nation because the ancestors have undergone a 'psychic trauma and a torture', consisting of a cerebral excitation touching them in the same way as the collectivity, and have created a mental engram in the form of a material imprint.[12]

Thus, employing windy assertions of physical substance in an attempt to cover up his total lack of empirical substance, an intellectual assures us that the ethnic character of a people is biologically transmitted. And if ethnic character as a whole is heritable, why not the propensity to compose verse in decasyllables? Is that any more improbable than the proposition, popular among the Bosnian Serb militiamen on the hills overlooking Sarajevo, that the city is the aggressor and its besiegers are the victims? Like Pavao Rudan, Striković offers a mechanism by which history leaves its mark on biology. But under the all-consuming imperative of blind loyalty, it becomes irrelevant whether an intellectual is constructing falsifiable hypotheses, or spouting mumbo-jumbo.

Surrounded by the war hysteria of Belgrade, Ivan Čolović has published an Orwellian satire in which he quotes a brochure for a fictitious book entitled *A Pure Serbia – New Vade-mecum of Eugenics*, published by the imaginary Military Institute of Ethnohygiene, promising 'everything you need to know about Serbian ethnic hygiene'.[13] 'The authors of this indispensable book are eminent ethnogeneticists, ethnopathologists, craniologists, anthropogeographers, migrationists, sappers, aviators and artillerymen.' Its contents are in the spirit of the genuine quotations Čolović appends; from contemporary Serbian nationalists, on the 'mystery of blood'; from Marinetti's Futurist Manifesto, on 'war, the only hygiene of the world'; and of course from Hitler's *Mein Kampf*, on racial purity.

Čolović's various examples establish that biologistic race concepts are part of the Serbian nationalist imagination. Biljana Plavšić's remarks about the genetic defectiveness of Bosnian Muslims are particularly reminiscent of Nazism. But the prospect of an Orwellian eugenic apparatus exists only in the realm of parody. It implies a coherent totalitarian ideology which simply does not exist in the post-communist world. The techniques of totalitarianism may be extensively employed, from control of the media to 'spontaneous demonstrations', but they are part of an array of options which can be adopted, combined or discarded as the situation appears to demand. The result is an oxymoronic kind of regime, in which totalitarian methods co-exist with liberal ones. Like Stalin and Hitler, President Slobodan Milošević of Serbia controls his people through his control of the media. Unlike the great dictators, however, he has permitted the existence of a few dissident organs, and of small liberal dissident groups like Čolović's Belgrade Circle. Race science is just one of a number of discourses which are pressed into service at various points. Yet there is nothing comparable to the totalising Nazi vision of politics as 'applied biology'. Biological race concepts are less powerful than hitherto, but more insidious.

The political potential of genuine science such as Pavao Rudan's studies becomes apparent when it is compared with the pseudo-science of a Plavšić or a Striković. As scientific discourse, it exists within a framework that allows the investigation of at least part of its claims to truth. That framework also crosses national and political boundaries, albeit unevenly. As a reflection on politics, human biological diversity studies may encourage an appreciation of complexity: of the heterogeneity within an ethnic group, or of the distinction between genetic inheritance and the sense of ethnic identity.

They may also make more or less explicit political observations. The Djerdap study, for example, refers to 'generations which freed themselves without prejudice from "ethnic purity" '. The region, the researchers suggest, shows evidence of longstanding 'ethnic symbiosis', which is 'only one of a series of changes which, both in the social and biological context, refreshes human communities and opens the way for new ethnobiological laws'.[14] Whatever these might be, and however elliptical the remarks, it seems clear that the authors are favouring the idea of diversity, and opposing it to the suspect notion of purity. Change refreshes; purity is associated with prejudice.

The more that biological diversity research interweaves itself with history, geography and other disciplines, the more it involves itself with the essential ingredients of the Romantic race concept. These integrated studies have the potential to appropriate blood and soil, very belatedly, on behalf of the Enlightenment. They may assert that, yes, forests and mountains may shape a people, but no, there is nothing about those influences which – in theory at any rate – is beyond the scope of rational inquiry. Conversely, such studies may be influenced by Romantic ideas, or may be appropriated by Romantic causes. Or, more likely, they may incorporate a variety of such influences and inclinations.

A case in point is a collection entitled *Genetic and Population Studies in Wales*, edited by Peter S. Harper and Eric Sunderland, and published by the University of Wales Press in 1986. It arose from the work discussed in the two preceding chapters on the genetics of Welsh Gypsies, in which Harper collaborated with E. Mair Williams; one of the accounts discussed earlier is included in the volume.

The first chapter, written by E.G. Bowen and published after his death, considers the relationship of land and people. It points out that, roughly speaking, Wales is a rectangle with two salient features, the peninsulas at the northern and southern ends of the country. These, Bowen asserts, were vital in bringing Wales its 'Mediterranean' population. He goes on to point out that Wales has no natural focus where a central authority could develop: draw diagonals across the rectangle and their intersection is roughly where the major valley ways arise, but it is also near the peak of Pumlumon. Instead, the accessibility of the southern plain left it open to Anglo-Saxons and Normans, who were then able to penetrate into the Welsh heartland along the valleys which open onto the lowlands. This resulted in the 'settlement of strong Alpine and Nordic elements'. The high moorland of the interior provided refuge for aboriginal types: early physical anthropological studies indicated a close similarity between the skulls of people in the Pumlumon area and the Palaeolithic specimens found in France and Spain.

Much later, with the industrial revolution, the railways brought migrants from other parts of Britain to the coalfields bordering the northern and southern shores. This increased the divergence between a heterogeneous, economically dynamic, English-speaking periphery and a core that was homogeneous but declining. The opposition between Inner and Outer Wales is the fundamental principle of the book, which is chiefly concerned with 'Welsh' Wales and the nature of 'Welshness'.

In this, as Eric Sunderland acknowledges in his overview, it stands squarely in the tradition established by Herbert John Fleure, the founding father of Welsh biological anthropology. In 1904, Fleure arrived at the University College in Aberystwyth, where, aptly and significantly, he later occupied a chair in geography and anthropology. He was a socialist, and he criticised the racist political movements that arose between the world wars. But his socialism seems to have been of the William Morris variety, rooted in a romantic view of his native Guernsey as an ideal egalitarian and co-operative community. Its virtues, he felt, were the product of mankind's natural instincts; he mistrusted the possibility of human progress through conflict.[15] The militancy of the South Wales industrial workers, numbers of whom adhered to a Marxist view of class struggle, must have been profoundly alien to him.

He certainly regarded the industrial populations of the principality as alien to Wales. Fleure did not object to the concept of race, only to what he saw as its political misuse. (Similarly, he believed in the principles of eugenics, but objected to the class bias he perceived in the Eugenics Society.) He sought out his research subjects in the rural regions, interested only in the identification and description of the 'real' Welsh. This population, pure and traditional, might perhaps be expected to preserve the naturally benign instincts of humanity in an uncorrupted state.

At the heart of Wales, on the high moorland of the Pumlumon region, Fleure detected what he believed to be physical similarities between living people and the Palaeolithic remains found in France and Spain. These, according to one contemporary popular account, included 'heavy brow-bridges [sic], low and sloping foreheads, massive and retreating lower jaw, all characteristic of Neanderthal man'.[16] In a chapter on ABO blood group frequencies among the 'indigenous population', I. Morgan Watkin, one of the doyens of Welsh genetic studies, notes that B levels are consistently raised in moorland areas. They are at their highest right in the centre of the country. Interpreting this as evidence of ancientness, Watkin also refers to a Scottish study which found that the B gene is much more common in the vicinity of megalithic monuments.

He also suggests that the traces of 'ancient mariners' may remain, in the form of an ABO distribution similar to those found in various Mediterranean island – the 'White Mediterraneans' of North Africa – and in sundry locations from Iceland to the Caucasus. This interpretation perpetuates the notion of a Mediterranean dimension to Welshness, once taken for granted as part of the tripartite European race-concept – Mediterranean, Alpine, Nordic – to which Bowen alludes in his opening chapter. Elsewhere, however, this race concept has succumbed to a combination of political and theoretical revisionism. Politically, they are inherently suspect, while more recent techniques of cranial analysis result in divisions which do not correspond to the traditional scheme, causing the Alpine class to disappear altogether.[17] Science once invested skulls with its fullest confidence; in the post-war era, the notion that peoples with similar crania must be related is apt to invite derision. 'There is no reason to believe that a "Mediterranean", "Dinaric", or "East Baltic" subject in Wales will possess the same residual heredity as the similarly typed individual in Warsaw,' observed Stanley M. Garn in 1962, echoing the comments of Tadeusz Bielicki that were noted in Chapter 2. 'Faced with "Australoids" marching over the South Pole to Patagonia, "Semites" migrating to Papua, "Koreans" to the Kalahari

. . . most of us gave up the typological approach in sheer exhaustion.'[18]

Speaking of the Inner Welsh as a whole, Fleure noted 'a remarkable persistence of type'. Convinced of the fundamental stability of physical characters, he used them as the foundation of his race concept, and was unable to accept Franz Boas's demonstration that skull form could be plastic.[19] Thanks in large measure to Boas's work, blood groups came to be regarded as replacements for bones, rather than complements to them. Morgan Watkin, however, sought to investigate whether the ABO system supported the earlier findings of physical anthropology. He concludes that, broadly speaking, it does. Other contributions also seek to incorporate more modern techniques into the Fleurian tradition. R.L.H. Dennis affirms earlier anthropometry using computers and updated statistical methods. As Eric Sunderland notes, Fleure's influence has induced a number of researchers to base their studies on the opposition between Inner and Outer Wales, including Sunderland himself.

This preoccupation with the 'real' Welsh obviously resonates with the idea of the 'true Gypsies': E. Mair Williams's decision to exclude Irish Travellers from her research population can thus be seen as consistent with the Fleurian tradition. Within the conceptual framework of the book, it is not regarded as problematic – indeed, Sunderland highlights her work as a demonstration that Gypsies are not an entirely isolated population, but have many links to both the Welsh and the English. Yet, as the last chapter made clear, Thomas Acton argued that Williams's conceptual scheme derived from the outmoded beliefs of scientific racism.

There are certainly critical elements in the volume, most notably in M.T. Smith's account of a blood group survey on the island of Anglesey. In this, the first blood group study to examine systems other than ABO and Rhesus in the north of Inner Wales, Smith questions some of the assumptions on which Fleurian research has been based. He argues that Anglesey too has a long history of migration, in both directions, and suggests that some sections of society may be more prone to migration than others. In order to accept Smith's argument that the population samples may be biased, it is not necessary to go along with him as he flirts with the idea that this might be genetically influenced, or that blood groups vary between social classes. The most striking instance of bias arises from the Fleurian tradition's dependence on surnames as indicators of kinship and descent, which excludes married women from the population samples.

Overall, the tone of *Genetic and Population Studies in Wales* is expressed by Frances Lynch's contribution, in which she alludes to the doubts now abroad concerning the traditional view of population movement as an endless series of invasions. The debate has centred on the so-called Beaker Folk, known for their round barrows and round heads. In 1977, a British archaeologist named Stephen Shennan advanced the argument that the beakers and associated objects were status symbols: the Beaker Folk were a class rather than a race.

Lynch herself seems to lean towards the traditional explanation, noting the association between round heads and Beaker graves in the Rhineland and France as well as in England. Her account of archaeology and fossil remains also alludes to the people who lived in the territory before the advent of farming; they were probably short, with long narrow heads and marked brow ridges: 'We may guess that their colouring was dark.' A fair guess, given the preponderance of darkness among extant humans, but here its effect is to imply a primordial quality to the association between the land of Wales and the short, dark, narrow-headed people of the Welsh stereotype.

Harper and Sunderland's collection contrasts markedly with *Archaeology and Language*, written by Colin Renfrew, who is a Professor and Master of Jesus College, Cambridge. Though Cardiff and Cambridge are only two hundred miles apart, and the books appeared within a year of each other, they appear to belong to different conceptual continents. In his discussion of the Beaker Folk controversy, Renfrew barely mentions the evidence of the round skulls, except to dismiss it. The reason is clear from his introduction: 'Craniometry . . . has in recent years enjoyed about as much prestige in scientific circles as phrenology . . . Racial anthropology – *Rassenkunde* to use the German term – has been convincingly discredited.'

Renfrew acknowledges that newer techniques of measurement

are being developed, but makes it clear that he doubts their usefulness. He is more positive about genetics, and indeed has subsequently prophesied a 'brilliant future' for DNA studies, looking forward to a 'new synthesis' of archaeology, linguistics and genetics.[20] A touch ironically for an archaeologist, his writing imparts a powerful message of progress: the new is good and the old is bad. All but the most recent osteometry is tainted by racism; the 'classical' genetic markers such as blood groups are better; most interesting of all are the new molecular techniques which examine DNA directly. Also progressive is the shift from biological to social explanations, epitomised by Shennan's argument that the Beaker Folk were not an ethnic group, but elite members of various groups. The underlying theme is of a movement away from race and racism.

The Cardiff volume conveys a message of conservatism that is equally powerful. It is not reactionary; it does not consider contemporary ideas and rail against them. But its presiding spirit is that of respect for tradition, and the classically conservative inclination to build upon what already exists, rather than to demolish and start afresh.

Metropolitan sensibilities may favour the explanation that these people are simply provincial and out of touch. This argument can be rejected on the grounds of its shallowness as well as its snobbery. The book's character derives from a commitment which goes deeper than science. It is prefaced by a quotation from the poet R.S. Thomas, held in awe not just for his talent but for the intransigence of his conservative nationalism. This is a conservative Welsh nationalist text, proud of its concern for ethnic purity and the terrain in which that purity may still be found; it is respectful not only of the elders who contribute chapters to the work, but also of the ideas that formed them in turn.

As is clear from the accompanying maps, showing the area where most people can speak Welsh, Inner Wales still extends over much of the principality. To analyse and record it is to help preserve it; even where the preservative effect is not as obvious as in the chapter 'Welshness and Disease', which considers the illnesses to which the 'real' Welsh are predisposed. There is a commonality of interests between this tradition of science and that of the culturally conservative strain in Welsh nationalism. Both are Romantically concerned with the relation between a remote land and an original people; both gaze inward, away from the urban and heterogeneous towards the rural and the homogeneous, towards the past.

In the period since *Genetic and Population Studies in Wales* appeared, the contrast between Romantic cultural nationalism and mainstream political nationalism has been heightened by the development of transnational European institutions. Political activists of the smaller nations of Europe increasingly see European integration as the best hope for fulfilling their national ideals. At a pragmatic level, the lesson has been driven home by large disbursements of European Community regional funds to peripheral areas. At a theoretical one, the framework of the European Union has been welcomed as the solution to the great traditional objection to independence, that small nations are incapable of standing alone. Nowadays intellectuals from Scotland to Slovenia talk a new language of nationalism, an 'inclusive' variety which understands national well-being to depend on interconnection with the rest of the world, not isolation. In this new perspective, it becomes easier to see interweaving – of industries, of culture, of genes – as the means by which identity is developed, not destroyed.

Yet Romanticism is incurable. Militant groups inspired by atavistic visions are a perennial feature of the nationalist fringe. The Scottish National Party was the unwilling host to a faction called *Siol nan Gaidheal*; Seed of the Gael, which sought to discover the Celtic warriors within unemployed urban youths, and dress them in paramilitary uniform. Meibion Glyndwr, the Sons of Glendower, became famous for burning down Inner Welsh houses bought by outsiders as holiday homes. This tactic emphasised the class dimension of the issue, and sympathy for the property-owners was diminished by the widespread feeling that houses should fulfil local needs rather than the taste of the distant rich for luxuries. But the essentials of the conflict are clearer in those cases where permanent residents have been warned to leave. Both in Scotland and Wales, local people have taken to referring to incomers as 'white settlers', a post-colonial term which implies an attitude of superiority on the part of the settlers, and an identification with the oppressed on that of the locals. Ironic as the expression is, it may be used to make the point that for the Seed and Son strain of nationalism, this is at bottom a matter of race.

Heritage is scarcely the exclusive preserve of nationalism, however. When the BBC began work with Professor Steve Jones of University College, London, on a major series about genetics, they did not adopt the title of his book, *The Language of the Genes*, even though its ideas are the basis of the series. The BBC's working title for it was *Origins*, because everybody is fascinated by where they came from. Although the uses of genetic diversity are numerous, the popular imagination responds to the one that deals with ancestry.

A similar sensibility was evident in an earlier documentary about genetic diversity, 'Sir Walter's Journey', broadcast in 1994 as part of the *Horizon* series. Sir Walter Bodmer's journey takes him from London to Pembrokeshire, to Cumbria, and finally to the Orkney Isles, north of Scotland and beyond the Celtic fringe. It also takes him back in time, his quest being to find genetic traces of people more ancient than the Ancient Britons. We are fascinated by aboriginal people even if they are not our direct ancestors.

The film, produced by Tim Haines, is unusually charming for a science documentary. It deploys a whole bag of cinematic tricks to evoke, very successfully, the various forms of memory and the ways in which they combine in our imaginations. To introduce the notion of genetic inheritance, we are shown old ciné footage of a younger Sir Walter and members of his family, his commentary making the reassuringly familiar point that relatives tend to resemble each other. This is the footage that all ideal families should have, and for that reason we are touched by other people's home movies, once they have acquired the patina of age.

The ciné device is used throughout the film, the black and white sequences giving the journey the aura of a childhood holiday, and Sir Walter's colleagues the look of a family. This is entirely authentic in the cases of two of them, his wife Julia and daughter Helen. At the level of a shared culture, the idea of a knight on a quest across the ancient places of Britain has an obvious romantic resonance – even if he does it in a Range Rover. The sights he witnesses on route – a stately home in Wales; traditional sports such as Cumberland wrestling and a kind of mob football in the Orkneys called the Ba' – speak to the sense of national heritage that has developed so greatly in recent years. These signs of tradition, of collective memory, are contrasted with the homogenisation represented by motorway service stations.

Before Sir Walter leaves the metropolis, however, he has to attend to some pressing business. He must establish that the study of genetic diversity need not be tainted by racism, and indeed may oppose it. He is seen in a Greek Orthodox church speaking to Father Andreas and his son Father Constantine, two Cypriot priests. Father Andreas reveals that he lost two other sons to thalassaemia, the hereditary blood disease which resembles sickle cell anaemia, but affects Mediterranean rather than black African people. Sir Walter tells his hosts that different types of mutation cause the disease in different areas. Greek and Turkish Cypriots share a thalassaemia variant with each other, but not with people from Greece or Turkey.

'Couldn't you prove biologically that we are descended from Ancient Greeks?' asks Father Constantine. Sir Walter replies that the common mutation implies a common descent for all Cypriots, from an indigenous population who were there before the Greeks: 'From a biological point of view, you are one people.'

This is received with wry amusement. 'It is news to us!' observes Father Andreas. 'Maybe you can solve our political problems as well,' suggests Father Constantine. 'I hope so,' answers Sir Walter, 'because it's so common that people have a common biological heritage, and yet it's the cultural difference on top that causes them to have the conflict.' He seems to imply that biological truth is more fundamental than cultural truth, not just to biologists but to society as a whole.

It is not at all surprising that the Ancient Greek question was uppermost in Father Constantine's mind. Greece's heritage is, naturally enough, a source of infinite pride; by the same token, Greeks are vulnerable to invidious comparisons between the accomplishments of Ancient Greece and the undistinguished record of the modern state. In a pseudo-academic variation on the commonest form of insult in the world, Greek-baiters of an especially malicious bent are prone to question the modern Greeks' ancestry.

A sustained effort on these lines appeared as an anonymous 'profile' of the Greeks in the London *Sunday Telegraph*, during a period in which Greece occupied the presidency of the European Community: 'The world must nod dumbly at the proposition that in the veins of the modern Greek, with his dark glasses, car-phone and phantom olive groves attracting EC subsidies, there courses the blood of Achilles.'[21] To sustain the heavy hand of invective, the essay went back to the 1830s to cite the opinion of the Austrian historian Jacob Philip Fallmerayer that modern Greeks contain 'not a drop of pure Greek blood'; further on, it claimed that 'the Greeks today are a mixture of Slavs, Turks, Greeks, Bulgars, Albanians, Vlachs, Jews and Gipsies'. The following week, a trenchant rebuttal from the Greek ambassador appeared on the letters page, pointing out that Fallmerayer was not the only Austrian to have been inspired by a similarly Romantic race concept. Rejecting any racial basis for Greekness, he drew upon the ancient authority of Isokrates, who proclaimed that 'we consider Greeks those who partake in our culture'.

Wisely, Sir Walter has avoided this Balkan quagmire. The first stop on his progress around Britain is in Wales's southern peninsula. Pembrokeshire is crossed by the Landsker line, marked on the ground only by the ruins of fortifications, but enduring in the division of the county into the Welsh-speaking villages of the north, and a southern region known as 'Little England beyond Wales'. Its character derives from Norman settlement; Sir Walter discusses this in Picton Castle, a Norman edifice, where he compares the appearance of the present occupant to the ancestral portraits on display.

The main point he wishes to make in this region concerns not the Normans, but the Flemings. These people were used by the English monarchy as a mobile human buffer, most of them being transferred from the Tweed by Henry I in 1108. They had helped keep out the Scots; now they were to perform the same function against the Welsh. (The Serbs of Krajina have a similar heritage.) Today, the Norman legacy is visible, as shots of Picton Castle's Norman arches illustrate. Sir Walter argues, however, that conventional history is unable to detect the traces of the Flemings, which persist only in gene frequencies. History is therefore incomplete without genetics. The images of the castle and the ancestral portraits suggest that genes must be readmitted to our notion of heritage, already so densely packed with historical baggage.

The case for readmission turns on whether the race concept will not inevitably slip back in with the genes. When science is popularised, it naturally begins to merge with everyday speech and ideas. In Cumberland, Sir Walter searches for Norse influence, visiting a primary school along with a pair of men dressed in Viking costumes, who demonstrate the use of spear and shield. Taking his lead from research by the Newcastle-based population geneticist Derek F. Roberts, which discovered children who bore a closer genetic resemblance to Norwegians than to their English neighbours, Sir Walter asks the class if any of them are of Viking stock. 'There's a nice blond-headed lad,' he observes. Within a matter of minutes, the putative Norse ancestors have been translated into fair-haired warriors. The Nordic type appears to be raising his blond head.

Sir Walter, of course, does not see it like that. But the difference between a convenient generalisation and the notion of type may be less clear to a lay person than it is to a scientist. In attempting to communicate with lay audiences, population geneticists may inadvertently revive ideas that remain latent in popular notions of human variation. Similar processes may also develop as mediators such as journalists or television producers become involved. They will inevitably be less tightly bound by scientific rigour, and will be more inclined to draw upon non-scientific traditions in order to tell their stories.

So far, population genetics has yet to capture the imagination of the media in a major way. But 'Sir Walter's Journey' is not the only indication that the idea is beginning to be grasped. One curious example appeared in *Le Figaro* magazine (the publication which began as a vehicle for the *Nouvelle Droite* guru Alain de Benoist, as described in Chapter 8). The cover of the 25 September 1993 issue featured the face of a man with hair and beard caked in mud. The headline read: 'Exclusive! We have met the Etruscans'. Inspired by the work of Alberto Piazza, who co-wrote the *History and Geography of Human Genes* with Paolo Menotti and Luca Cavalli-Sforza, the magazine gave an archaeologically inspired makeover to some of the 1,815 inhabitants of Murlo, a Tuscan village south of Siena.

Piazza and his colleagues had previously mapped the frequencies of 34 genes, all governing characteristics of blood. Using the statistical technique of principal components analysis, so important to IQ studies, they produced a series of genetic maps of Italy. The one based on the second principal component, representing 18 per cent of the total variation, has a round genetic eye right in the middle. This, the researchers noted, corresponds to the area occupied by the Etruscans before 800BC. Etruria subsequently expanded and flourished, but vanished as the Roman Empire rose. The end of the story, for pupils in Latin classes of more recent times, was that nobody really knows what became of the Etruscan people. But they did enjoy a revival in mythical form during the Enlightenment, when the cult of Etruscomania, or *etruscheria*, claimed that Etruscans had matched the achievements of Greek civilisation on Italian soil.[22]

The Italian geneticists offered a new kind of answer to the question of what happened to the Etruscans: they are alive and well and living in Tuscany. According to the headline of a *Newsweek* report of Piazza's research, a quote from Cavalli-Sforza, 'The Genes tell the Whole Story'.[23]

To investigate the Etruscan question further, Piazza settled on Murlo because it was a geographically isolated village. Modern culture transcends geography, of course; in the summer of 1993, the local cinema was showing *The Last of the Mohicans* to the putative last of the Etruscans. *Le Figaro* emphasised how the people of the village were culturally indistinguishable from other late-twentieth-century Italians.

Professor Piazza appears in the magazine himself, achieving the remarkable feat of looking benign while holding a vial of blood in one hand and a skull in the other. Fifty years ago, such a pose might have symbolised the old and the new markers of human variation. Now, with the advance of molecular biology, bones serve as a source of ancient DNA. Sir Walter Bodmer's team took a similar approach in the Orkneys, comparing blood samples from living Orcadians with a few strands of DNA that his collaborator, Erika Hagelberg, had succeeded in extracting from 5,000-year-old bones found at a burial site known as the Tomb of the Eagles. Both studies are part of a co-ordinated European research effort. Alberto Piazza chairs a network of twenty-five laboratories around Europe, which have received some funds from the European Union's Human Capital and Mobility Programme, under the title 'Biological History of European Populations'; Sir Walter Bodmer chairs the European Regional Committee of the global Human Genome Diversity Project, which operates under the auspices of the Human Genome Organisation.

The interview with Piazza contained the standard disclaimer, in the form of a quote unequivocally rejecting both the notion of racial hierarchy and the notion of race itself. But this statement was little more prominent than the admission, buried within excited tabloid prose, that the research was still in progress – and even when complete, could not actually prove a link between the people of Murlo and the Etruscans.

Overwhelmed by a heady rush of Etruscomania, however, the magazine had made up its mind. Several pages of glossy colour photographs showed various residents of Murlo dressed in styles suggested by Etruscan art, with which they were juxtaposed.

The centrefold featured Aldo Teccioli, a pizza and pasta restaurateur with a figure to match, reclining voluptuously in a bathrobe and a gold chain. 'Sometimes, the resemblance is hallucinatory,' the caption claimed. Actually, Signor Teccioli was the only model who bore a really strong resemblance to his prototype, in this case a statue from a sarcophagus. But the text uses his example to claim that the physical similarities between ancients and moderns are 'indisputable'. Had he been alive to read it, Herbert Fleure's heart would have been gladdened.

The moral here is not that such remarks reveal a secret racist agenda, smuggled in under the guise of non-racist population genetics. It is that even the most forthright of anti-racist disclaimers is not sufficient to control the full range of ideas that will be sent into circulation by this research discipline and its ramifications in the media. Some of these ideas, particularly the vaguer and more peripheral ones, will derive from, or speak to, older traditions of racial science. Some will reassert notions that disappeared along with the old race science, but will place them in a new conceptual framework.

No system will succeed in clearly separating old from new, folklore from science, myth from data; especially not in Europe, the fabulous continent of endless identities which are infinitely recombinant but largely incommensurable. Macedonia is not an aberration, just a particularly acute instance of incommensurable identities happening to occupy the same space at the same time. As an aggregate of ideas, Europe is only tenuously connected to any scientifically measurable reality. Hans Magnus Enzensberger expressed that truth when he appropriated the Shakespearian reference to the 'seacoast of Bohemia', via a poem by Ingeborg Bachmann, for the title of an essay about the future of Europe.[24] Although a medieval Bohemian ruler did briefly secure a toehold on the Adriatic shore, the essential impossibility of a maritime Bohemia allows the phrase to serve as a metaphor for the distance between reality and European national fantasies.

In the stabler regions of Europe, any new sense of biological heritage is likely to lead to theme parks rather than concentration camps. It is quite easy to imagine places like Murlo in which the gene frequencies of the inhabitants, suitably framed by costumes and tableaux, become tourist attractions. The people of Murlo and Kirkwall, the Orcadian capital, were united in their willingness to provide blood samples and their fascination with the findings. Keen to know where they came from and what makes them special, they helped science without payment (though the mayor of Murlo did hand out commemorative certificates). In the future, other towns and villages may grasp the commercial potential.

The poor prospects for a Fourth Reich should not obscure the fact that the data have political potential too. Piazza and his colleagues allude to a connection between population structure and politics in the *History*. They refer to the finding, by H. Le Bras and E. Todd, that family structure and political opinions are highly correlated. In north-eastern France, nuclear families prevail, and parties oriented towards the free market have tended to record their highest votes. The north-west, Le Bras and Todd say, is a stronghold both of patriarchal extended families, where Father's word is law, and authoritarian parties. Social democratic tendencies are strongest in the south-west, home of the benevolent, non-patriarchal extended family.

France, like Gaul before it, is thus divided into three parts; Le Bras and Todd suggest that this division may be extremely old. The Italian geneticists endorse the idea, arguing that family structure is probably one of the most conservative aspects of culture, even antedating the language spoken. In fact, they suggest, it may be as old as the genes. They point out that the map drawn up by Le Bras and Todd to show the distribution of the socialist vote is similar to their own map of gene frequencies, which illustrates the division between Palaeolithic and Neolithic stocks in France. But they are clear that, although there would probably be a high correlation between socialist votes and the O blood group, the latter does not cause the former. O genes are simply a neutral indicator of population origins, and political tendencies arise from the different cultural traditions which have prevailed in these populations.

Such a position rediscovers the idea dear to the heart of traditional race theory, that a population may have such a thing as a character. The success of the Liberal Party in Yorkshire, for example, could then be ascribed to short-headed Asiatic influence.[25] By locating this phenomenon in the cultural sphere, however, it maintains the firebreak between modern population genetics and political controversy. Behaviour geneticists have no such inhibitions; indeed, they seem constitutionally disposed to polemic. Although none of them are likely to assert that O genes encourage voting for socialists, a link between genes and political views has become a favourite claim among behaviour geneticists, drawing upon comparisons of twins raised apart. It is probably only a matter of time before some psychologist devises an evolutionary theory of political outlook. Meanwhile, in Japan, a folk genetic tradition maintains a widespread popular belief in the association between blood groups and personality, to be considered when forming intimate relationships and creating balanced teams in the workplace. It appears to have originated with a publication in 1916 asserting a link between academic success and type A, the commonest in Japan; the work was a reaction to contemporary Western claims of superiority for European blood groups.[26]

Cavalli-Sforza, Menotti and Piazza do not make any observations about political opinion in their own country. Yet they are working at a time in which the differences between the populations of Italy are so marked that the continued existence of the Italian state is in question. Piazza and colleagues depicted their principal components analysis, in a paper entitled 'A Genetic History of Italy', using a series of computer-generated colour maps.[27] The first principal component, depicting the largest part of the variation in orange and red, revealed a distinctive area in the south which corresponded to *Magna Graecia*, the area dominated by the Greeks eight centuries before Christ. The second, the one which brings out the Etruscan eye, was rendered in shades of blue. The most notable feature of the third is a dark region in the north. The fourth map is a synthesis of the first three, in which Italy is divided roughly into northern, central and southern bands.

These do not correspond to features of antiquity, but they do bear a resemblance to contemporary political divisions. Elections held in 1994 showed a tripartite division in which the south was dominated by the self-styled post-fascists, the former communists maintained a belt of support in the centre, and the north fell to the right-wing populists of the Northern League. The ideologue of the League, Professor Gianfranco Miglio, dwells on the distinctive characters of the peoples of Italy. Drawing upon the same fount of antique myth as *Le Figaro* magazine, he embellishes his vision of northern glory by equating the region with Etruria.[28] Southerners, he has said, are 'anthropologically' different.[29] The geneticists' maps illustrate his point.

Miglio, who prefers to locate the difference in climate rather than racial origins, might not feel inclined to draw upon such evidence himself, but when both scientists and society at large are preoccupied with ethnic diversity, such connections will undoubtedly be made. As it happens, there is a precedent in this instance. Another Northern League, the one founded by Roger Pearson, claimed that southern Italians are descended in part from African slaves imported by the

Romans, whose miscegenation with them set in train the decline of their civilisation.[30]

So far, population geneticists have been content to address the political implications of their work by rehearsing the standard scientific anti-racist line. They argue that, by recording more details of the bewildering complexity of human variation, they can reinforce the scientific message of the meaninglessness of race. That has certainly been a key theme of the grand debate about race over the past half-century, which has been part of a still grander debate about human universals. But the newer era of nationalism and ethnic self-assertiveness has raised new political issues.

As far as the study of genetic diversity in Europe is concerned, these may not matter very much. If the initial ventures in Murlo and Kirkwall are anything to go by, European communities are likely to see such exercises as an interesting, even inspiring, exploration of their cultural heritage. They will be happy to have their ancestry examined within a scientific paradigm, confident that they and the investigators share broadly the same European culture. When scientists from the wealthy quarters of the world attempt to conduct studies among poor and isolated peoples, however, the reaction is likely to be very different. The lines of conflict are already clear, in the wake of an acrimonious controversy over the Human Genome Diversity Project.

12

Through the Bottleneck

THROUGHOUT ITS HISTORY, one of the most important guiding principles of anthropology has been to hurry while stocks last. In the nineteenth century, anthropologists realised that their subject material was crumbling under the pressure of the same colonising processes that made their study possible. Now that physical anthropology has been reborn as a form of molecular biology, it is the population geneticists who are haunted by the fear that they are racing against time. The forces of technology and the global economy are against them.

In the early 1970s, referring to the transformation of Britain by late twentieth-century mobility, Derek Roberts wrote that, if genetic variation exists, 'it is important to record it now while there is still time'.[1] In 1991, Luca Cavalli-Sforza, Allan Wilson and several other geneticists warned: 'The genetic diversity of people now living harbors the clues to the evolution of our species, but the gate to preserve those clues is closing rapidly.' Their two-page statement was entitled 'Call for a Worldwide Survey of Human Genetic Diversity: a Vanishing Opportunity for the Human Genome Project'.[2] The data gathered would speak to the concerns of history, archaeology, linguistics and anthropology; they would weave genetics into those discourses. 'It would be tragically ironic if, during the same decade that biological tools for understanding our species were created, major opportunities for applying them were squandered,' the paper observed.

To a significant extent, this document defined what was to become known as the Human Genome Diversity Project for some time to come. The roots of the controversy in which the project became embroiled two years later can be traced to the perspective which dominated it, that of Luca Cavalli-Sforza. He and Wilson were the leading figures in the initiative, but there were major theoretical differences between them. Cavalli-Sforza wanted to concentrate on isolated populations. He argued that information on origins could be maximised by concentrating on groups that were well defined and clearly isolated, at least until recently. Cavalli-Sforza favoured the pinpoint, Wilson the net. Wilson wanted to impose a grid upon the map, and sample the aboriginal people living at each node of the grid, whoever they might be.

Cavalli-Sforza believed that there were not enough aboriginal populations to fill up the grid, and that ethnic affiliations had to be taken into account. But his own strategy was open to the criticism that it biased the results. At the project's first planning workshop, Mark Stoneking, a member of the Wilson school, objected that a survey based on well-defined ethnic and linguistic groups would inevitably find that humanity is composed of well-defined ethnic and linguistic groups.[3]

There was also the question of what such groups should be called. Cavalli-Sforza was labelled a 'colonialist' for describing European populations as 'ethnic groups' and African ones as tribes. The implication of the criticism is that he considers European ethnic organisation to take a higher form than the African. Rejecting the charge, Cavalli-Sforza observed that 'the comment reveals a taste for quibbling, and ignores many differences in the social structures most frequent in these two continents, which must be taken into account in a sampling programme'.[4]

The degree of difference in social structures is, however, a moot point. In a paper presented to a meeting on the HGD Project sponsored by the Wenner-Gren Foundation for Anthropological Research, Alan Swedlund noted a revisionist trend in anthropology, developed by Morton Fried in the 1970s, which argued that, rather than being a stage of social organisation on the road to statehood, tribes were structures largely imposed by colonial states, and that the interbreeding patterns of non-European peoples were not nearly as neat as anthropologists had made out.[5] If this is true, it undermines the opposition between the 'modern' population structure of Europe and the 'traditional' societies found outside it. It also undermines the rationale for sampling the genes of 'well-defined' groups. They may not be as clearly defined in real life as they are in Ph.D theses.

The preference for the pinpoint over the net also affected the meaning of the project. There are two grand ideas within the concept of human genetic diversity. One is that of diversity *per se*; the other is that of origins. If the primary purpose of the survey is to map the extent and richness of human biological variation, then Allan Wilson's grid plan would be an appropriate starting point. Conversely, if the project's main concern is not with what we humans have become, in all our bewildering hybridity, but with what we once were, it will naturally focus on 'isolates of historic interest'. Or on races, according to the definition offered by the geneticist Gunnar Dahlberg in the early 1940s: 'a race is an isolate or a collection of isolates'.[6]

The pinpoint will also be a lot cheaper. The costs of processing a single sample were estimated at around $2,300. There are thought to be between four and eight thousand populations around the world that ideally ought to be surveyed. If two hundred samples were to be taken from each of these, the budget would reach three or four billion dollars. The funds initially available to the project, provided by the US National Institutes of Health and other bodies to cover the costs of a series of planning workshops, amounted to a mere $150,000. It seemed to make sense to go for a more modest survey, involving perhaps a few hundred populations. Since some of these groups were on the verge of disappearance, it also made sense to get their samples in the bag first. And an appeal for funds always gains from a sense of urgency.

Sir Walter Bodmer, addressing a Western lay audience in a popular book, said that the result of the project would be a 'living museum of human diversity'. It would 'rescue' samples from 'all the world's threatened peoples before their unique stock is lost or mixed with Western genes'.[7] When Bodmer speaks to the general public, his observations affirm a vision of science as a thrilling cultural venture which is fundamentally benign and humanistic. In Western eyes, this 'molecular Noah's Ark' may appear to be in the same spirit. But to other audiences, it looks like a pernicious variety of genetic reductionism, which saves molecules instead of saving people. For endangered human groups, a 'living' race gallery, in which the part is taken for the whole, is no better than the old race galleries of skulls from around the world.

Rhetoric like this strengthens suspicions about the West's apparent desire to keep indigenous peoples untainted by Western genes or Western culture. Once the natives start watching television, they cease to be interesting. They are felt to have let themselves down, to have lost their authenticity. The West's sympathy appears conditional upon the natives' resistance to Westernisation. The Ark metaphor, moreover, suggests an unfortunate equation of indigenous humans with animals.

The heart of the matter is the distinction between human diversity and the variety of nature as a whole. One may cherish an 'unspoiled' wilderness, but then lose interest in the area once it has been 'spoiled' by having a town built upon it. But people remain people after they modernise and interbreed. They may well occasion far more cause for humanitarian concern than they did in their previous state. Yet the rich seem readier to attribute value to the otherness of indigenous peoples than to acknowledge the bonds of common humanity.

When dealing with indigenous peoples, professionals from the wealthy world nowadays have to contend with the volatile politics of the post-colonial era. Among the most prominent actors on the stage are activist groups, which were particularly energetic in the sphere of conferences and debates in 1993, the UN Year of Indigenous Peoples. During this period they came to a view on the HGD Project. It was unanimous and vehement. The International Conference of Indigenous Peoples, held in Guatamala City, resolved 'to categorically reject and condemn the Human Genome Diversity Project as it applies to our rights, lives and dignity'.[8]

The line was laid down by RAFI, the Rural Advancement Foundation International, from its Ottawa office. Having expressed criticisms to representatives of the nascent HGD Project, RAFI went public in May 1993, circulating the correspondence to like-minded groups around the world. In a memorandum entitled "Immortalizing" the (Good?) Samaritan', which made heavy rhetorical weather of the fact that the 480-strong Samaritan community of Israel were among the peoples suggested for sampling, RAFI denounced the HGD Project as a 'human scavenger hunt' that was 'only out for blood'.

The substance of RAFI's critique was a combination of pre-existing concerns about the exploitation of materials to which indigenous peoples lay claim, and an anti-Western ideology which regards science as an arm of its capitalist enemy. RAFI made the general objection that $23 million, a figure suggested as the rough cost of a realistically sized survey, would be better spent on the immediate welfare of the populations in question. It also noted that at $2,300, the estimated cost of processing a single sample, was more than the per capita GNP of 110 nations.

RAFI's most dramatic argument was the suggestion that 'knowledge of an indigenous communities' [sic] unique genetic make-up makes it theoretically possible for unscrupulous parties to devise cheap and targeted biological weapons effective against specific human communities'. This idea seems to be based on two observations about modern military technology. One is that biological weapons are cheap – which they are, in comparison to nuclear ones. The second is that of the trend towards precise targeting. Here, this serves as a cue for the reappearance of the notion of racial essence.

The bomb's-eye views from 'smart' weapons in the Gulf War are easier to grasp than the notion of clines in gene frequencies. To the public, they are true science fiction, encouraging the belief that technology may be capable of anything in future. 'How soon will it be before they start altering these genetic materials and make movies like *Jurassic Park* come true?' demanded Victoria Tauli-Corpuz of the Cordillera Peoples' Alliance in the Philippines. 'Who will stop scientists from developing super human beings or monsters, or weapons for biological warfare from these genes?'[9]

The fact that *Jurassic Park* dealt with the re-creation of dinosaurs suggests that her questions should be read primarily as a declaration against science and in favour of its social control. They also illustrate the extent to which the lay public engages with science through science fiction, which approaches the laws of nature with dramatic licence. The manipulation of human genetic material undoubtedly raises extremely important questions about the social control of technology, but these relate to human genetics as a whole, not just the sub-discipline of diversity studies. When it comes to the targeted ethnic weapon, however, the question is not so much who stops the scientists, but what?

The military advantage of a race bomb would be that it would not need to be pointed in the right direction, like a gun, nor would it be hazardous to the attackers if they became exposed to it. To meet the latter criterion, it would have to lock on to characteristics which reliably differentiate friend from foe. In a war between the Italians and the Japanese, to take a hypothetical example, any participant lacking the active form of the aldehyde dehydrogenase enzyme could safely be assumed to be on the Japanese side. Whether that or any other difference could be used as the basis of a weapon is another matter.

Moreover, the maximum efficiency of an anti-Japanese race bomb based on aldehyde dehydrogenase would be about 50 per cent, since that is roughly the proportion of East Asians with the inactive form of the enzyme. It is true that many weapons systems are even less efficient – but only in practice, not theory. On the other hand, there is a proven method of biological warfare with a theoretical maximum casualty rate of 100 per cent, a cost approximating to zero, and – given an isolated target – no risk of harm to friendly elements. It entails dropping a dead animal down a well.

Some populations, particularly isolated ones in which genetic drift has taken place, contain 'private polymorphisms' not found in any other groups. Many more such polymorphisms will undoubtedly be brought to light by the HGD Project. But these variations may only appear in a small percentage of the population. Ethnic groups are not defined by exclusive genes, shared by all their members. Unless the entire body of knowledge developed through the past half-century of population genetics turns out to be wrong, there is no racial essence.

One of the HGD Project's premium selling points is precisely that, by developing this understanding, it will make the world a better place: 'Most importantly . . . the results of the Project are expected to undermine the popular belief that there are clearly defined races, to contribute to the elimination of racism and to make a major contribution to the understanding of the nature of differences between individuals and between populations.'[10] If that is the case, researchers associated with the project will have to work rather harder on the public understanding of science than Dr Rene Herrera of Florida International University, who tried to arrange a sampling visit with Chief Irvince Auguiste of the Dominica Carib Council in the summer of 1993. In a letter confirming telephone discussions, Herrera wrote that 'the purpose of our visit is to obtain blood samples from members of your Carib Nation that are known to be pure . . . and not contaminated with other races'.[11]

In the event, the Caribs were swayed by the rhetoric of the Cordillera Peoples' Alliance, which had arrived trailing endorsements by allied groups from New Zealand to Tunisia. The rhetoric was reproduced in local newspaper articles, including one by Julius Corriette, a Carib Councillor, who embellished it with the term 'genetic prostitution'.[12] A couple of weeks before the Florida team were due to arrive, the council postponed the visit.[13]

It seems that although the science of human biological diversity speaks the language of non-racism and anti-racism, adhering to the concepts of population genetics, it is wont to lapse into older dialects when it addresses lay audiences. And those dialects are not confined to the political right. The left-wing ideology of indigenous peoples' activism appears to be compatible with ideas of racial essentialism, and therefore of racial purity. This is suggested particularly strongly by remarks made by John Liddle, director of the Central Australian Aboriginal Congress, about what the CAAC dubbed the 'Vampire' Project: 'In the last 200 years, non-Aboriginal people have taken our land, our language, and our culture, our health, even our children: apparently now they want to take the genetic material that makes us Aboriginal people.'[14] The phrasing suggests that the Aboriginal identity is essentially a biological one, with culture and history forming a sort of superstructure.

To guard against the kind of theft that Liddle denounces, RAFI drew heavily upon its experience of campaigning on the issue of plant biodiversity. The basic dynamics appeared very similar. Plant genetic material associated with indigenous populations has been collected and taken North for storage, where it is available for commercial exploitation. In the HGD Project's early planning discussions, the commercial potential of the samples gathered was assumed to be negligible. Recent developments elsewhere induced RAFI to think otherwise. United States law permits the *de facto* patenting of genes or gene products – technically, the patent applies to the processing of the material. Internationally, a network of patent culture depositories has been established for the storage of biological materials. The largest in the world is the American Type Culture Collection at Rockville, Maryland; in 1991, it contained nearly 18,000 deposits.[15]

Several individual cases seemed to set ominous precedents. In 1976, cancerous cells removed from John Moore, a Californian leukaemia patient, were converted into a cell line and, eight years later, patented. Moore sued, claiming a share of the profits. After another six years, in 1990, the California Supreme Court rejected this claim. Once removed from his body, the court ruled, the cells no longer belonged to him. Sir Walter Bodmer firmly agrees with the decision. We shed genetic material all the time, he points out.[16] This is not calculated to allay the suspicion that if a population group refused to co-operate with the HGD Project, sampling could be conducted clandestinely from sources such as barbershop floors.

The ATCC catalogue already contains an entry for a T cell line derived from a member of the Guaymi people of Panama, complete with patent claim infringement warning and price code. And the possible commercial potential of local gene mutations has been illustrated by the case of Limone, an Italian village whose residents enjoy the benefits of a gene which results in lowered levels of high density lipoprotein, and therefore lowered levels of heart disease. Patents on the AI-Milano gene are held by Kabi Pharmacia, a Swedish company which has also secured the services of Cesare Sirtori, the gene's discoverer.

RAFI succeeded in turning the Guaymi case into a *cause célèbre*, campaigning with the Guaymi General Congress for the withdrawal of the patent claim and the return of the cells to Panama. Isidro Acosta, president of the congress, declared that patenting human

DNA 'violates the integrity of life itself, and our deepest sense of morality'. On the material plane, he established that human genetic material is not excluded from – though not specifically included in – the General Agreement on Tariffs and Trade.

The same month, in October 1993, Green Members of the European Parliament tabled an emergency motion whose demands included a halt to the patent claims and to the HGD Project. The US claim was withdrawn the following month; the month after that, the controversy reached a climax at the World Council of Indigenous Peoples conference in Guatemala City. The conference's collective ire came down on the head of Hank Greely, professor at Stanford Law School and chair of the ethics subcommittee of the HGD Project's North American Committee, during four hours of acrimonious debate. 'We won't allow them to exterminate our race!' cried one delegate. A second feared that sociologists and anthropologists 'will study how we make love'. A third suspected that the CIA was at the bottom of it.[17]

By this stage, it was easy to overlook the fact that Pat Roy Mooney, executive director of RAFI, had previously expressed the hope that scientists and indigenous peoples might eventually co-operate on a project whose principal goals 'deserve support'.[18] Even before the HGD Project controversy, many indigenous peoples' groups had grown uncooperative towards researchers, tired of servicing the Ph.D industry with no tangible benefit to themselves. In recent years, they had developed a sense of grievance over the exploitation by Northern capital of plant biodiversity, and now they were faced with the prospect of their own biodiversity being turned into profit far away.

They wanted, above all, to assert their ability to act. The draft report of the second HGD Project planning workshop explained that, in the absence of any fully acceptable term, it referred to groups 'in danger of physical extinction or of disruption as integral genetic units' as ' "Isolates of Historic Interest" (IHI's), because they represent groups that should be sampled before they disappear as integral units so that their role in human history can be preserved'.[19] The Aboriginal Health Council of South Australia commented: 'This rhetoric constructs indigenous people as irrelevant to the present or to *history-making*. The impression conveyed is that . . . they are mere anachronistic curiosities of an ancient past fit only for scientific reflection – rather than people engaged in shaping the futures of the nations to which they belong.'[20] Showing a more critical approach to race concepts than the Central Australian Aboriginal Congress, it also questioned the notion of 'integral units', scenting a trace of the old idea of 'full blood' that had been applied to Aboriginal ethnic identity in former times.

In sum, the indigenous peoples' groups mustered an array of criticisms that ranged from the constructive to the abusive, and from the astute to the frankly ludicrous. Although they were united in their opposition to the project as it stood in 1993, there appeared to be underlying differences between those who regarded the HGD Project as a colonialist crime against nature, and those who would be happy to see it proceed, given an acceptable balance of power between science and indigenous peoples. At the centre of the alliance sat RAFI, which had maintained the latter position while doing so much to whip up the fears underlying the former. The letter of its line endorsed co-operation; the spirit encouraged confrontation. Hank Greely, principal target of the flak, calls it 'slash and burn politics'.

Much of the acrimony might have been avoided if discussions on the project had begun a year or two earlier. As it was, the incipient HGD Project collided with the conference round of the Year of Indigenous Peoples before it had evolved very far beyond the original 'Call for a Worldwide Survey of Human Genetic Diversity'. Of the two leading figures, Cavalli-Sforza and Wilson, only one was able to continue to participate directly: Allan Wilson had died in July 1991, around the time that the 'Call' was published. His followers continued to advance his arguments, however, and as more scientists became involved in the discussions, the emphasis shifted away from Cavalli-Sforza's isolates. In addition to theoretical considerations, other scientists felt that the threats to small population groups were not as great as had originally been claimed. Another factor was the European Community funding, which helped the European affiliates of the HGD Project to become the first actually to start collecting samples. Although relationships between the rich North and the poor South will remain a fundamental factor in the project's operations, it has become a more rounded enterprise. It no longer concentrates quite so strongly upon the South's acutest sensitivities.

The influence of the controversy upon the project is debatable. Hank Greely acknowledges that the exchanges with RAFI on patenting were useful, but the possibility of a continuing dialogue seems to have been severely limited by the manner in which RAFI made its case.[21] 'Quite frankly,' Pat Roy Mooney had told Greely at one point, 'your phone call, letter, and memo . . . renders the Ethics Committee quite incredible'.[22] The feeling appears to be mutual. But whatever credibility is accorded to RAFI and the four or five allied groups which voiced their opposition, their comments have a wider resonance. Adopting a popular electronic bulletin board form, members of the project's North American Committee produced a set of answers to 'Frequently Asked Questions', including the ones about cloning and ethnic weapons. The kind of questions posed by the activist groups seem quite typical of lay reactions to the linkage that the HGD Project makes between ethnicity and genes.

It was not until late in 1994 that the project finally issued a formal statement of aims and principles.[23] The document was based on the results of the final planning workshop, held the previous year in Sardinia, home of the second most outlying set of gene frequencies (after the Lapps, or Saami) in Europe. The principal aim of the project was defined as being to 'investigate the variation occurring in the human genome by studying samples collected from populations that are representative of all of the world's peoples'. Its main value was said to lie 'in its enormous potential for illuminating our understanding of human history and identity'. Thus the two grand ideas of diversity and history were synthesised.

Three other areas of value were identified. The data would provide information on genetic factors in disease. The connections made between geneticists and members of other disciplines – biologists, linguists, historians, anthropologists and archaeologists – would create a 'unique bridge between science and the humanities'. Finally, it would help to combat racism. With this document, the project finally grasped its own potential, which lies in its ability to make organic connections between different domains of knowledge.

'The primary case . . . for the Human Genome Diversity Project is cultural,' it declared.

That being so, ethical and legal considerations were all the more important. At the outset, Cavalli-Sforza and his colleagues had worried mainly about what would happen during sampling, such as the question of making sure the subjects gave informed consent, or of respecting local customs, or the problem of what to do about medical needs that became apparent to the researchers as they went about their work. Issues of intellectual property rights received more attention in the 1994 document, but they were not resolved. The project itself was not intended to make a profit, but one of its principles was that the material it gathered would be available to any researcher. Referring to the exploitation of indigenous people's knowledge of plant biodiversity, the proposed guidelines said that any patent 'should include provision for the financial return on sales to benefit the sampled population or individual'. This is a very long way from the kind of control over the project that the activists demanded for indigenous peoples. Liberal goodwill among scientists had yet to be fully translated into firm guarantees. Even if it had, it would have stopped far short of allowing real authority to pass from science to subjects. And many of the subjects might not even accept such a transfer if it were offered, since they reject the authority of science in the first place. As noted in Chapter 7, the Lakota Indians do not accept the scientific worldview, which conflicts with their creation myth. The two camps cannot simply agree to differ, since they are involved in a protracted dispute over the 'repatriation' of American Indian remains held in museums (the term is enshrined in the Native Americans Graves Protection and Repatriation Act of 1990). In some cases, however, the origins of the remains are unknown. As archaeologist Steven Shackley notes, they may have no more lineal connection to the present-day Native American occupants of a region than to the museum curators. 'Allowing one set of people to dictate the treatment of relics that belong to another runs directly contrary to the ideal that motivated NAGPRA in the first place,' he argues.[24]

One method used by scientists to try to determine the ethnic affinities of 'unclaimed' remains is cranial analysis. Bronco Lebeau,

Repatriations Officer of the Lakota, is outraged. 'Where do they get off trying to tell us who we are or are not related to?' he demands.[25] The Lakota use ceremony, not analysis, to address these questions of self-determination.

Similar issues have arisen in Australia, where fossils from Kow Swamp were handed over to local Aborigines and reburied in 1990, although they represent people who are clearly distinguishable from living Australians by their extreme robustness, and their scientific interest lies in precisely that fact.[26] The Human Genome Diversity Project may well become involved in this kind of power struggle, with both symbolic and material dimensions, particularly if genetic analysis is perceived by its subjects as a threat to the connection they claim to the lands they inhabit. Both spiritual identity and material resources may then be at stake.

It would be wrong, however, to assume that the project will be no more than a tug-of-war between metropolitan science and indigenous peoples. While isolated populations have remained at the heart of the project (since they allow researchers to reach back farther in time than those more open to the fluxes of migration), it now affirms that most living people are members of highly complex populations, and that it is 'designed to represent all our highly variable species, past and present'.

To do so, a range of sampling strategies will be required. These may well include one proposed by Fatimah Jackson, of the University of Maryland, for sampling African-Americans. Most of the movements of people to be investigated by the project were voluntary ones – such as the southward and eastward 'Bantu expansion' of West Africans, which started about 3,000 years ago, or the settlement of the Americas, the chronology of which is hotly disputed. Jackson, however, is concerned with the population created by the greatest involuntary population movement in history. In the scientific literature, periods in which populations are severely reduced are laconically described as 'bottlenecks'. Usually these are prehistoric (such as the one which appears to have accompanied the exit of modern humans from Africa) or peripheral (such as the sinking of a boat off the Atlantic island of Tristan da Cunha in 1885, with the loss of fifteen of the island's tiny population). When the bottleneck in question is the Middle Passage of the slave trade, and the conditions endured by its survivors, the moral and political legacies are as enduring as the genes.

Among the most striking features of African-American health are the high levels of hypertension – about double the incidence in whites – and kidney disease. A genetic predisposition has been suggested, but has been greeted with suspicion in many quarters. It is tainted by the race science tradition. Richard Cooper and Charles Rotimi point out that high incidences of hypertension are found in Russia and Finland, but have not led to anybody to speculate about a genetic predisposition in those populations.[27] It appears to be another way of implying racial inferiority: it smacks of blaming the victim, rather than the social inequalities which are consistently associated with poor health. Like the IQ argument, it can be used to justify neglect: why bother to improve health services if the patients are doomed by their genotypes anyway?

Cooper and Rotimi point out that the association between blacks and hypertension is less consistent outside the United States. Data from the Caribbean are contradictory; in South Africa, rates are low in the countryside and high in the townships. Within the United States, Cooper and Rotimi suggest, research may be confounded by environmental risk factors such as obesity, lack of potassium, or stress. They discuss the possibility that, in the United States, the incidence of hypertension may be elevated by 'the unique effect of psychosocial stress experienced by people of color in Euro-American society', or by subtler effects of racism. The difficulty, they acknowledge, is that such factors are virtually untestable. As for the evidence of physiological differences between blacks and whites, which include higher intracellular sodium concentrations in blacks, as well as differences in half a dozen other metabolic systems, the authors consider that the very diversity of these variations makes the desire to work them into a unified theory 'surprising'. An evolutionary selection hypothesis would be 'uncomfortable'. The message is one of deep reservations about attaching meaning to genetic diversity.

A prerequisite for the adequate testing of any genetic hypothesis is the measurement of the range of genetic variation throughout the African-American population as a whole. A dataset describing a normal population is derived from Caucasians, as in America it will tend to be; a population displaying different gene frequencies is likely to be viewed as abnormal. This applies not only to black Americans, but to non-Europeans in general. As the HGD Project argues, a set of reference databases for different populations will reduce the bias of biomedical science towards people of European descent. The benefits of such databases would include more efficient matching of tissue types for organ transplants, and a better understanding of the different mutations that produce genetic disorders in different populations.

The tendency to regard African-Americans as abnormal is apparent in one formulation of a genetic hypertension hypothesis, which refers to a 'renal defect in the handling of sodium'. This suggestion is based on the evidence that the sodium of dietary salt may raise blood pressure. In this respect, black Americans are more sensitive than white ones. They seem to retain salt in their bodies longer; they have higher blood pressure than both white Americans and black Africans. The suggested mechanism is that, when a quantity of sodium enters the body, it raises blood pressure, which stimulates the kidney to increase the rate at which sodium is excreted, until the rate at which the ion is entering the body equals that at which it is leaving. However, in people disposed to retain salt, the blood pressure remains elevated.

Over the last twenty years, a number of scientists have speculated about the possibility that salt metabolism in sub-Saharan Africans might have adapted to the climate. A substantial amount of salt is lost in sweat, so there might be an evolutionary advantage to a mutation which reduced the rate at which salt was excreted. In the 1980s, attention turned to the Middle Passage, in which thousands of slaves were transported across the Atlantic. Under conditions of close confinement, it was observed, they would have been subjected to intense heat, and probably deprived of salt in their food. Clarence E. Grim then pointed out that huge amounts of salt are lost through diarrhoea and similar diseases, like the 'bloody flux' or dysentery, and vomiting. Seasickness, sweat and stools would have combined to evacuate the salt from slaves' bodies.[28]

The slave trade may have been profitable, but it was monstrously inefficient. According to the figures presented by Grim and his colleague Thomas W. Wilson, about twelve million Africans were transported to the Americas from the sixteenth to the nineteenth centuries, but about a third of those captured never reached the New World. About one in ten died before the march to the coast reached its destination; about 12 per cent died awaiting embarkation; and 12–15 per cent succumbed during the voyage. Of the survivors, 10–30 per cent were dead within three years. Planters calculated the cost of slave mortality rather as motor traders calculate depreciation on cars today. After seven years or so, the cost was written off altogether.

The next generation fared even worse. The mortality rate among the infant offspring of slaves reached 50 per cent, and so the importation of fresh stocks from Africa had to continue. This was the nature of the slavery 'bottleneck'.

According to the 'Grim hypothesis', the Middle Passage in particular exerted a heavy selection pressure for salt retention. The prospects of survival were significantly higher for those individuals whose genetic constitution caused them to retain salt. They passed on those genes, which became widespread in the African-American population. The selection pressure for salt retention would have been maintained by the sweat of forced labour, endemic diarrhoeal disease, and cholera epidemics. Nowadays, however, Americans eat a diet rich in salt, which has a greater effect on the blood pressure of black Americans than whites. Wilson and Grim point out the inadequacy of the term 'defect' to describe the presumptive genetic difference between the groups. If their hypothesis is correct, the capacity to retain salt conferred an advantage in one environment, but has become a disadvantage in another.

A general difficulty with evolutionary hypotheses is that they tend to be much easier to generate than to test, and the Grim hypothesis has been criticised on these grounds. Wilson and Grim refute the charge, suggesting several possible lines of inquiry. One is to exploit the bureaucracy of the slave trade, comparing mortality figures with the records of salt supply. Among living peoples, they propose surveys of blood pressure, and its sensitivity to salt, that compare

Africans and African-Americans; they also urge a search in other parts of the world for peoples with predominantly low blood pressure.

The Grim hypothesis has a formidable symbolic potential. As well as proposing that the legacy of slavery is encoded in the genes, and is killing black people more than a hundred and thirty years after emancipation, it points to a new concept of African-American identity.

For around a century, race classification in the United States was organised around the 'one-drop' rule, maintaining the purity of the white category by classifying anybody with the slightest trace of African descent as black. At the same time, scientific racism dwelled upon the fact that black Americans were a hybrid African-European stock. One of the favourite parting shots of the old guard, with the post-war tide flowing against them, was to observe that a number of black men who had achieved high distinction did not actually look very black. Their success was attributed to the size of the European portion of their genetic inheritance, rather than to the advantages arising from closer contact with white society. Martin Luther King, according to Henry Garrett in the *Mankind Quarterly*, was a 'complex intelligent mulatto'.[29]

The obsession with lightness was reproduced among blacks. At one time, the chic mulattoes of Washington would apply the 'paper bag test' to their prospective party guests. Anybody darker than a brown paper bag did not receive an invitation. Today, the one-drop principle has been appropriated for the cause of ethnic solidarity. It is contradicted, however, by an abiding sense of profound unease about the continuous gradation of African and European mixture to which light skins attest.

There can be few issues so fraught and complicated as that of sexual contact between black and white Americans. Without attempting the impossible task of removing the biological element of the complex, one or two observations about the social meaning of the genes may usefully be made. History makes it extremely difficult, not to say absurd, for black Americans to be content with the entire range of their genetic heritage. The European gene flow was in large part from master to slave, at best a relationship of extreme inequality, and at worst straightforward rape. Black experience in the New

World associated race-mixing with oppression, and that negative association was affirmed by the preference of race science for 'pure' stocks over hybrids. The Grim hypothesis opens up the possibility of a new concept of African-American identity, which escapes the stale opposition between the pure and the hybrid, and unites the biology of black Americans with their history.

As a moral tale, the Grim hypothesis is perhaps a little too neat. It has been subjected to trenchant criticism from several perspectives. For the historian Philip D. Curtin, it illustrates the inability of scientists to do history. They don't understand that historical papers become outdated just as scientific ones do; they fail to give page numbers when citing books. Current historical opinion, he says, puts the total number of slaves transported at around ten million, and the death toll of the Middle Passage at 10 per cent. While conceding that this makes more difference to the quality of Wilson and Grim's scholarship than to their hypothesis, he also queries the percentages they give for mortality between capture and embarkation; any such figures, he says, are guesses.[30]

Once in the Americas, Curtin points out, the slave populations fared differently in different regions. Death rates outstripped birth rates among slaves in the tropics well into the nineteenth century, whereas in North America slave populations were increasing rapidly by then: 'In fact, most historians of the American South now believe that the diet and disease patterns for slaves in the South was [sic] pretty much the same as those for the poor Whites.'

Venturing onto scientific turf, Curtin observes that 'it is a curious claim that seasickness, which rarely lasts as long as 48 hours, could affect one's descendants more than 200 years later'. But Wilson and Grim claimed that seasickness was one of several salt-depleting factors, which in aggregate had a significant effect on mortality. Oddly, Curtin does not question Wilson and Grim's claim that the most common recorded causes of death were diarrhoeal diseases – 'flux' – and fevers, but he dismisses diarrhoeal disease on the grounds that cholera was absent from the Atlantic basin, and diarrhoea kills infants, not adults. He does not explain what else he thinks might have killed the victims of 'flux'.

Curtin drew attention to the way that the slavery hypothesis was uncritically reported in the popular scientific press. For Cooper and Rotimi, 'its widespread currency demonstrates just how susceptible the scientific audience is to acceptance of genetic explanations when the subject is race'.

A third critique, published with Wilson and Grim's paper, rejects the hypothesis but endorses the idea of evolutionary selection that Cooper and Rotimi evidently find uncomfortable. Fatimah Jackson agrees that the Middle Passage may have exerted selection pressure for the retention of sodium and other electrolytes.[31] But she points out that the circumstances of the translocation to the Americans are likely to have stimulated genetic diversity. A number of studies suggest that extreme environmental stress can damage DNA itself. Normal repair mechanisms are overwhelmed, leading to outbreaks of genetic disorder in which mutations, recombinations and stray fragments of DNA multiply. The result is an increase in variation.

Comparisons between Jackson's paper and Curtin's suggest that they perceive slavery in very different ways. Citing eyewitness reports, Jackson writes of the 'potential for infection provided by the close proximity to diseased and dead fellow captives and to blood-and mucus-covered decks, the excessive heat generated by the tight packing of human cargo under conditions of clearly inadequate ventilation, the high rates of starvation (often self-imposed)'. By contrast, the sanguine Curtin considers 'we can assume that the slavers' self-interest was incentive enough to make them carry adequate salt and water – and food as well, though probably not of the best quality'.

At a theoretical level, however, the two appear to agree on the heterogeneity of conditions for slaves in different parts of the New World. Like Curtin, Jackson notes that, once in the Americas, the Africans found themselves in a variety of environments: the genetic implication is that they were therefore subject to different selection pressures. Interbreeding with Europeans became widespread, but at the same time many populations were isolated, encouraging genetic drift. All in all, Jackson considers it unlikely that the selection pressure of the Middle Passage would have been consistently maintained in the Americas. Her model is one of severe genetic constriction during the Atlantic crossing, followed by a flowering of diversity.

This emphasis on diversity is a characteristic of Jackson's work, as is her insistence on the need to overcome the conceptual division between biology and culture.[32] Both themes point to potentially important tendencies in science. The first is the affirmation of diversity as an inherently good thing. Many scientists, such as the sociobiologist Edward O. Wilson, have embraced the planetary ethic of commitment to the preservation of biodiversity.[33] The extension of this principle to biodiversity within our own species seems both natural and inevitable; the beginnings of such a process can be seen in the connections made by the indigenous peoples' activists between the politics of plant biodiversity and the HGD Project.

It is perfectly possible for scientists to affirm biodiversity solely on ethical or cultural grounds, using their scientific understanding to illustrate the beauty of diversity. They may further be inclined, however, to provide a scientific rationalisation for the ethical imperative. The favourite argument for preserving biodiversity – that it is an untapped medical resource of possibly enormous proportions – can easily be applied to human genetic variety. 'Variation is our species' wealth,' says Fatimah Jackson.

Besides its affiliation to contemporary Green ideas, the argument harks back to a different tradition. It proposes that steps be taken to maintain the genetic stock of the species. In other words, it is racial eugenics, but the race in question is the human race.

The second tendency is one that should be applauded by Richard Lynn, who emphasises the notion of 'gene-culture co-evolution' in his criticisms of environmental explanations for group differences in intelligence. Jackson is a scientist prepared to explore the relationship between genes, culture, and ethnicity, just as Lynn and other race-conscious scientists advocate. Yet in both her scientific and political outlook, she is radically opposed to the perspectives that psychologists like Lynn and J. Philippe Rushton promote.

This is hardly surprising in an African-American scientist with a biological orientation, of course. But the significance of her position is that, while politically and scientifically anti-racist, it does not rely on the denial of difference. Workers affiliated to the HGD Project acknowledge that there may well be times when the study of human biodiversity may produce uncomfortable findings. The view of these scholars is that, if differences exist, we just have to deal with them.

13

One Race, One Science

THE ASSOCIATION BETWEEN knowledge and joy, frequently strained in the course of scientific inquiry, finally reached breaking point with the advent of DNA sequencing techniques. Journals were filled with page after page of A's, G's, C's and T's, representing the four bases which bear the DNA code. It was an exercise in surveying, not interpretation. Fairly soon, the task was automated, leaving machines the job of reading ever more genetic sequences into the scientific record.

As the data generated by these unsleeping drudges mounts up, the need to impose form upon them will become ever more acute. The examples of anthropometry and psychometrics suggest that, once it is felt to be important to collect data of a particular kind, the volume of information soon encourages the development of new techniques for its organisation. At present, the clinal approach to population genetics yields a different map for each trait, like weather maps drawn on different days. They do not evoke the joy that comes with the revelation of an overarching pattern, but in this they serve a useful ideological purpose by demonstrating how difficult it is to sustain racial constructs beyond the level of external form.

To represent combinations of traits, Alberto Piazza and his colleagues have developed a method that relies upon computers and colour graphics. The analytical technique itself derives from Spearman, who first identified g, and the data themselves are based on a few dozen genetic markers. As the sequences accumulate, gene mapping spreads out from these landmarks to the form of the landscape as a whole. New representational models will be developed to interpret the data generated by the Human Genome Project and its successors. Given the rapidity both of genetic advance and social change, it seems unwise to deny the possibility that these models will produce new forms of ethnic classification, on which the race gallery of the twenty-first century may be based.

Whatever advances in complexity are achieved by genetic models, investigations of the human genome will still concentrate on certain limited areas of interest within the three billion bases of the human DNA sequence. Prominent among these will be regions thought to influence mental traits. The record so far is patchy, the typical story being the announcement by one team of a 'gene for' this trait or that, followed by the failure of other researchers to replicate the results. A group at the University of Texas identified a genetic sequence that they believed to be associated with alcoholism, for example; a review of the ensuing literature by workers at the National Institute on Alcohol Abuse and Alcoholism concluded that, with the Texas group's papers excluded, the balance of findings failed to uphold the association.[1]

Investigators like these will not be deterred by their failure to achieve immediate acceptance, however, and it is striking that even the preliminary work on alcoholism was racialised. The allele in question varied markedly in incidence between ethnic groups: Kenneth Blum, leader of the Texas researchers, argued that those in which it was most common were also those with the highest rates of alcoholism. There was one notable exception, the Japanese, half of whom have the allele. Blum argued that they have low rates of alcoholism because many also have a different gene that prevents them from metabolising alcohol. The inactive variant of the aldehyde dehydrogenase allele is believed to produce an aversive effect by allowing the accumulation of acetaldehyde from the breakdown of alcohol; the body reacts with an accelerated heartbeat, feelings of weakness and heat in the stomach, and facial flushing.[2]

Blum and his colleagues have gone on to identify a genetic sequence they believe underlies cocaine dependency, a phenomenon integral to the issue of black criminality as it is currently conceived.[3] Genetic markers associated with the neural receptors in question are, it has been suggested, differently distributed among blacks and whites.[4] And another group of researchers has claimed an association between the supposed alcoholism marker and 'criminal aggression'.[5] A report by the US National Academy of Sciences and the National Research Council called for more research on biological factors related to violent crime, as well as on whether 'male or black persons have a higher potential for violence than others, and if so, why?'[6] Genes, neurotransmitters and behaviours, the last of these racially marked and defined as pathological, are being steered into orbit around each other.

Far more momentous, will be the search for genes influencing cognitive ability. 'The isolation of the first gene involved in determining "intelligence" (whatever that is) will be a turning point in human history,' declares Benno Müller-Hill of the *Institut für Genetik* in Köln.[7] His immediate concern is with eugenics, and the possibility of a movement to weed out 'inferior' genes. But genes for genius, or at any rate high intelligence, have immense social significance too. In the elitarian tradition, from Galton to Murray, societies are shaped principally by their few exceptionally gifted members, and so the relative fortunes of different ethnic groups will be similarly determined. Identify a 'gene for' an IQ over 130, and the cognitive elite is identified as being different in kind from everybody else, rather than simply the upper end of a continuous spectrum. The quickest way to establish the mental inferiority of an ethnic group as an accepted truth rather than a disturbing contention is to label a stretch of DNA as the genius gene. No matter how mysterious, limited or roundabout its effect on intelligence quotients may be, the label will stick. And any group found to be short of this sequence will be labelled intellectually inferior as a whole.

At first glance, however, high intelligence is a straightforwardly positive trait, and so associated genetic inquiries are not as controversial as those in pursuit of biological bases for antisocial behaviour. Intelligence is a complex phenomenon, and even the most bullish hereditarians expect any single gene to explain no more than a few per cent of the variation within a population. If geneticists are going to choose a gene to pursue, therefore, it seems natural to be interested in one for genius rather than in one that underlies the cognitive abilities of the masses in the middle of the distribution. In 1991, the National Institute of Child Health and Human Development granted Robert Plomin $600,000 to fund a search for genes associated with high intelligence.[8]

This project relies upon *g*, as measured by intelligence quotients. But the inclination to look for interesting genes connected with cognitive ability might well encourage researchers to turn to notions of multiple intelligences, favoured by anti-Jensenists such as Stephen Jay Gould as an alternative to concepts of intelligence based on a single number. And they may in due course begin to feed reports into the public domain which, however qualified, affirm the idea that different races have their brains organised in different ways.

Cognitive ability, however, is a proper source of joy, and the racial implications of research into it have not been widely admitted, except where the investigations are explicitly organised around ethnic categories. This is true of genetic advances in general, which are recognised to pose profound ethical dilemmas, and to be potential sources of discrimination. Yet critics seem either reluctant to explore the ethnic dimensions of these issues, or simply oblivious to them – except where the racial implications are obvious, as in studies of crime focused upon the inner city. In his article 'The Shadow of Genetic Injustice', quoted above, Benno Müller-Hill observes that 'doubtless some ethnic differences will be found. Suddenly, genetic racial injustice will be a fact'. And that is all he has to say on the subject, in a two-page essay. He appears to take ethnic differences so much for granted that neither their reality nor their implications merit further discussion. Yet Müller-Hill of all people is aware of what may emerge from a political synthesis of race and biology: he is the author of *Murderous Science*, a study of the role of scientists under Nazism.[9]

Part of the reason for this blank area in the visual field is that many of those concerned with the ethical implications of genetics are working within a liberal paradigm which holds to an ideal of race-blindness. The current *Zeitgeist* encourages the liberal disposition to see issues in terms of individual choice. Will middle-class parents, for example, choose to abort foetuses which fail to carry genes predisposing for high intelligence? At the same time, the notion of race-blindness has come under sustained attack from more radical tendencies which advocate compensatory privilege for oppressed groups. This induces liberals to avoid mention of race altogether. They are thus predisposed not to notice it, and if they do, they are inclined to ignore it in the hope that it will go away.

What has actually happened is that, as Kenan Malik shows in his book *The Meaning of Race*, the race concept is as powerful as ever, but it has been translated from the domain of biology to that of culture.[10] As race disappeared from the physical anthropology textbooks, 'difference' became embedded in cultural theory. A liberal, universalising impulse combined with the race concept's internal flaws to minimise the presence of race in science, but the repressed returned as militant anti-racism. The Enlightenment, rather than the Romantic reaction against it, got the blame for the death camps; universalism was accused of being inherently racist. The Romantic doctrine of innate racial character was reborn as the idea of difference, entirely the product of culture, but just as immutable. In the world of absolutist relativism, all knowledge is subjective and specific to groups. African and European Americans live in different cognitive worlds; so do women and men. These worlds are radically incompatible, and so there is no point in trying to communicate between them. 'Difference . . . has been resolved into *indifference*, an unwillingness to engage with what anyone else has to say,' Malik comments.

Whatever the challenges posed by the radical notion of difference, its public salience does have the effect of affirming a more general and less confrontational notion of diversity. At a popular level, diversity is becoming understood as a quality that is inherently good. It is, in fact, identified as a proper cause for celebration. Fatimah Jackson's dictum, that 'variation is our species' wealth', can be seen as a statement of the connection between the idea that we should value the cultural diversity of society, and the imperative, asserted by Edward O. Wilson among others, that we must understand the importance of the diversity of life as a whole.[11]

Like ecological principles in general, Jackson's aphorism may be vulnerable to banalisation. It is certainly valid for a scientist to take pleasure in contemplating the sheer variety of nature, just as it is for anybody fortunate enough to be able to walk through a forest or swim above a coral reef. But to have meaning in a human social context, the principle has to be associated with an inquiry into just how, and why, genetic variation is valuable. The reward then arises not simply from contemplation, but from understanding. If the insights are sometimes disturbing, the fact that they have been obtained will remain a source of satisfaction in itself.

There are similar compensations, often grim, in textual diversity. 'As a card-carrying First Amendment (near) absolutist, I applaud the publication of unpopular views that some people consider danger-ous,' said Stephen Jay Gould, that Marxist bogeyman of hereditarian mythology, in his review of *The Bell Curve*.[12] And that may serve as the order of the day for the coming period, in which the post-war settlement on race and science may be revised. If Salman Rushdie should be free to explore ethnicity in his own way – and he certainly should – Jean Philippe Rushton should be free to portray race in his.

Equally, university authorities are entitled to dissociate themselves from the views of race scientists, since science is socially embedded and apt to be influenced by the wider beliefs of those practising it. And public bodies are not under any obligation to finance race science with public money. They are, however, obliged to consider the social effects that might result from such research. If it is proper to provide public funds for research projects which might benefit society by extending our knowledge of disease, it may be also be proper to refuse funding for projects which could harm society by exacerbating racial antagonism. Members of all ethnic groups pay taxes, after all: it would plainly be improper to spend the money on activities that might worsen the position of any group relative to others.

Freedom of speech also implies the right to protest of those outraged by scientific racism. If the protestors were themselves more inclined to recognise their targets' rights to free expression, however, we might be spared the Pavlovian cries of 'McCarthyism' and its attendant clichés, amid which the hereditarians so enthusiastically posture as martyrs.

What actually is racism, though? (Posing this question at the end of the book, rather than the beginning, is quite deliberate.) Under the post-war anti-racist order, science provides the guarantee for a definition which states that racism is a belief in the innate superiority or inferiority of ethnic groups. A broader definition, also within the terms of this order, identifies racism as any ascription of deeper significance to the physical traits associated with various populations.[17] If race is scientifically trivial or meaningless, it is easy to arrange all the possible aspects of racism within a single framework. There is no need to prove hostile intent, since all negative characterisations are by definition the products of prejudice.

But what if science came to accept that there are meaningful innate differences in mental capacity – which is the nub of the issue – between human groups? Then we would either reject science, or we would all become racists in the broad sense of the term. We could still condemn racism in the form of hatred, aggression, or the denial of legal rights, but the edifice of anti-racism would have been critically weakened.

However, the history of race science suggests that it can consistently be relied upon to express the *Zeitgeist* in which it operates. This implies that even if science did in the future restore to favour the idea of meaningful differences between populations, it might well reject it again under subsequent historical circumstances. For the time being, one approach to resolving the conceptual difficulties might be to split the concept of racism into two. The meaning of 'racism' could be restricted to racist violence, hatred, prejudice, discrimination on racial grounds, and indeed the issues denoted by the term 'institutional racism'. The belief in the objective existence of innate differences in mental capacity or character between populations might then be designated 'scientific racism'.

Its practitioners would thus be dissociated from malign intent, but at the same time have to accept a qualified characterisation as racist. That seems like a reasonable deal. And, given the race scientists' voluble concern for free speech, they should also appreciate the merits of plain speech.

Although history shows that race concepts are malleable, the lesson is limited within any given *Zeitgeist*. Scientists and lay citizens alike are profoundly influenced by the totality of beliefs that surround them. Although one may acknowledge their historical contingency, they may still create an irresistible impression that they are natural.

The dilemma is acutely apparent on what used to be the political left. Since the end of the 1970s, when the neoliberal boom began, all political ideologies based on substantial state ownership of industry have undergone collapse or drastic revision. The market paradigm now appears universal: the options for socialists are to work within it, or to cling to the old faith in the hope that they will live long enough to see the pendulum swing back. The true believers are not necessarily wrong just because they are in a minority, but they are doomed to political marginality for as long as the market paradigm remains in the ascendant.

There is, of course, substantial overlap between the political left and those who would prefer not to consider the role of biology in human affairs at all. Conversely, biological determinists are frequently enthusiastically right-wing. The case of the 'gay gene', however, defies the neat arrangement that associates nature with the right and nurture with the left. Many homosexual men welcomed reports first of a brain structure and then of a gene associated with homosexuality. They felt that the news affirmed what they have always known, that they were 'born that way', and that if this were accepted as scientific truth, their civil rights could not be denied.

Like other political movements that have replaced the socialist left, but remain recognisably tendencies of the left rather than the right, gay activism is based on the consciousness of difference, rather than a vision of human universals. It finds a natural model in the idea of the ethnic group: this is made explicit in the phrase 'Queer Nation'. In such a vision, the gene and the brain structure are markers of national identity.

By no means all gay men will endorse this vision, and many emphasise that sexuality can be seen as a continuum – just as scientists now see clines rather than discrete races. But the fact that many gay men are well disposed to biology, despite the gay history of being defined as pathological by medical science, offers an intriguing hint about the future of attitudes to biology on the terrain formerly occupied by the left. If gay men can see biology as a tool for liberation rather than oppression, then perhaps other groups – particularly those preoccupied with difference – may draw similar conclusions.[14]

Within science and its public ramifications, the position of anti-racism remains far stronger than that of the left in politics. Jensenist intelligence theory claims to be mainstream, but geneticists remain committed to public affirmation of anti-racist principles. A determined but discreet minority current in anthropology remains committed to the renewal of traditional race concepts; the great bulk of the discipline maintains the anti-racist line.

During the same period that has seen neoliberalism take over the political and economic world, its sociobiological cousin has consolidated itself within science. As president of the Behavior Genetics Association, Sandra Scarr felt able to claim that opposition in the 1980s was confined to a few 'eccentrics' such as Leon Kamin and Stephen Jay Gould.[15] Her own criticisms of the racial genetic hypothesis on intelligence illustrate the success of the evolutionary behavioural model. It now has its own right and left wings, with all parties agreeing on the basic terms of debate and investigation.

In the sociopolitical sphere, this period of triumph for the doctrine of the invisible hand has run a course from joy to gloom, from optimism to pessimism. It has happened before: in Victorian England, for example, when early faith in self-improvement turned, without any substantial change of ideology, but with the onset of a period of economic decline, into a defensive middle-class fear of the dangerous classes and of foreign threats. At the outset, the spirit of enterprise is assumed capable of cleaning up the ghetto; after a while, the spirit of enterprise comes to be regarded as a precious quality to be protected, along with those who embody it, from the underclass.

In such a climate, scientists working within an evolutionary-behavioural model may be expected to join the drift to pessimism. Vincent Sarich's insistence that nothing can be done about human nature will seem to ring truer than the claims of those who argue that we need to know our innate tendencies in order to guard against them. Pessimistic and deterministic ideas from science will then feed back into lay discourse, which will become more pessimistic, and will feed its pessimism back into science, and so on.

Within the political sphere, particularly in Europe, the racist right seems likely to continue to exert a strong influence upon the mainstream, with which it will to some extent merge. Whether it will make much use of biology remains to be seen, but if tradition counts for anything – and on the far right, it counts for a lot – it will do so if it can. That in turn depends upon whether biological accounts of difference attain any degree of acceptability in political discourse.

And that may have a considerable effect upon mainstream conservatism, legitimating not just cuts in education or welfare spending, but a general conviction that those at the bottom of the heap are literally useless. Old-fashioned socialist eugenicists believed in public spending to get the best out of all sections of the population. National efficiency demanded that if a person's proper station was to be a hewer of wood or a drawer of water, then he or she should be a fit and healthy hewer or drawer. Nowadays hewing and drawing are increasingly done by machines, leaving a surplus population. This has the appearance of inefficiency, but any question of responsibility for the problem is resolved by the conclusion that the surplus is genetically incapable of playing any useful part in modern, cognitively loaded society.

A wild card in any consideration of the future is the decline of popular faith in science. It was the power of the belief in scientific modernity that permitted the emergence of a political movement that saw itself as applied biology. Now science is just one more belief system among which to pick and choose. This makes the effects of developments in scientific thought less predictable. It suggests that people will be less impressed by scientific pronouncements on race. On the other hand, it also suggests that they will be less concerned whether such pronouncements are dominant or those of a minority opinion. They will pick and choose whatever suits them.

There may also be a fashion effect. Certain scientific ideas are more readily adopted by the lay world than others, bestowing on them the impression of a degree of influence they may not actually enjoy in their original domain. One of the most popular recent admissions from science to the culture at large has been the notion of chaos. Up to now, sociobiologists and the managerial classes have been talking the same language, of game theory, rational choice, inputs and outputs. The central idea of chaos is that minor perturbations may have major effects: a butterfly's wingbeats may cause a storm. If that is adopted as an explanatory model, then similarly distant and minute agencies might be seen as a more plausible reason than IQ averages for, say, Europe's sudden winning streak of the past half millennium. In short, it may come to be understood as a mathematical explanation for chance.

The major obstacle between race science and the public remains the great historical separation between biological and cultural levels of explanation. Several scholars cited in this book share a commitment to oppressed ethnic groups, and point towards a reintegration of the two great streams of knowledge. They are not the first to have such a vision:

> History itself is a real art of natural history – of nature developing into man. Natural science will in time incorporate into itself the science of man, just as the science of man will incorporate into itself natural science: there will be one science.[16].

So at last we seem to have some common ground, uniting Fatimah Jackson, the Romani scholar Thomas Acton, Richard Lynn, Jean Philippe Rushton, and the author of these words, Karl Marx. When hardline race scientists like Lynn and Rushton talk of gene-culture co-evolution and hope that it will blossom into a paradigm, they seem to be looking forward to a victory similar to that of capitalism over socialism. But if a new paradigm does emerge, it will embrace a far greater diversity of opinion and mindsets than evolutionary behavioural science does at present.

And the growth of diversity may be a powerful counterbalance to the growth of pessimism. Some will engage with evolutionary biology and genetics precisely because they realise that the research will be done anyway, and they do not wish hardline hereditarians – or racists – to have a monopoly on the interpretation of the data. Not everybody will see culture as the glove-puppet and genes as the hand inside. Not everybody will hear the voice of the genes in every utterance of culture. It may be possible to negotiate an articulation between the sciences, a unification which allows each to remain independent where appropriate.

Many positive rewards start to appear in such a vision. There are many genuine joys to be experienced. Prominent among them is the ability to relate language, gene frequencies, bodily traits and history together, which allows one to reconcile the pleasures of romanticism and rationalism. It may even become possible to learn how to take an authentic pleasure in the biological aspects of human diversity.

Or it may not. In the balance of cultural forces, a pessimistic and selfish *Zeitgeist* might prevail over the attempts of liberals to build a kinder, gentler paradigm. Their efforts might amount only to sugaring the pill. The temptations of reductionism will always be present in any way of thinking that incorporates biology, and the trenchant environmentalist arguments of the 1970s and 1980s stand as reminders of biological determinism's characteristic flaws. Nevertheless, it no longer seems sufficient to respond to race science with a reassertion of a binary opposition between biology and society, as implied in the rallying-cry of an appeal distributed on the Internet in the wake of *The Bell Curve*: ' "RACE" IS A SOCIAL CONSTRUCTION, NOT A BIOLOGICAL CONCEPT.'[17]

It is true that the only certain race is the human race. Perhaps, however, the time has come to explore how biological variation and social constructions are related. Dealing with difference may be easier said than done. But denial no longer appears to be an option.

There are new developments in this book's field every week – news items, new publications, new ideas. To help make *The Race Gallery* a continuing discussion, and to further the principle of democratic, pluralistic discourse that underlies the book, *Race Gallery* pages will be maintained on the World Wide Web of the Internet. They can be found at:

http://www.hrc.wmin.ac.uk/racegallery/

Notes and References

I RASSENSAAL

1 Nancy L. Stepan, *The Idea of Race in Science*, Macmillan, London, 1982.

2 Carleton Coon, *The Races of Europe*, Macmillan, London, 1939.

3 Tacitus, *The Agricola and The Germania*, translated and introduced by H. Mattingly, translation revised by S. A. Handford, Penguin, Harmondsworth, Middlesex, 1970. Quote in parentheses is as given by J.S. Huxley and A.C. Haddon in *We Europeans*, Jonathan Cape, London, 1935.

4 *Profil*, Vienna, 8 March 1993.

5 Ibid., 6 September 1993.

6 *Nature* 364, 26 August 1993.

7 *Sunday Telegraph*, London, 19 September 1993.

8 Erhard Busek, Answers to Written Parliamentary Questions 5383/J-NR/1993, Bundesministerium für Wissenschaft und Forschung, Vienna, 19 November 1993; *Profil*, Vienna, 13 December 1993.

9 Johann Szilvássy, pers. comm., 23 November 1994.

10 F. W. Rösing and E.–M. Winkler, 'Zur Paradigmengeschichte der Taxonomie: von Bernier über von Eickstedt zu Hiernaux', *Homo* 43(1), 1992, 29–42.

11 Ivan Bernasovský, *Seroanthropology of Roms (Gypsies)*, Košice, 1994; Joseph Deniker, *Les races et les peuples de la Terre*, Paris, 1926.

12 Božina M. Ivanović, 'Morphological Properties of Yugoslav Population Groups', in *Etnoantropološki Problemi*, Monografije, Knjiga 11, Department of Ethnology, Faculty of Philosophy, Belgrade, 1990.

13 Živko Mikić, *Antropološka Struktura Stanovistva Srbije (Contribution à la Structure Anthropologique des Habitants de la Serbie)*, Belgrade, 1988.

14 Mark Snyderman and Stanley Rothman, *The IQ Controversy: The Media and Public Policy*, Transaction, New Brunswick, New Jersey, 1988.

15 Sharon M. Lee, 'Racial Classifications in the US Census: 1880–1990', *Ethnic and Racial Studies* 16(1), 1993, 75–94.

16 See for example Lawrence Wright, 'One Drop Of Blood', *New Yorker*, 25 July 1994; *Newsweek*, 13 February 1995.

17 Julian S. Huxley and Alfred C. Haddon, *We Europeans, A Survey of 'Racial' Problems, With a Chapter On Europe Overseas, by A.M. Carr Saunders*, Jonathan Cape, London, 1935. Charles Seligman and Charles Singer also collaborated in the work: see Elazar Barkan, *The Retreat of Scientific Racism: Changing Concepts of Race in Britain and the United States between the World Wars*, Cambridge University Press, Cambridge, 1992.

18 Pierre van den Berghe, 'Racism', in *The Encyclopedia of Cultural Anthropology*, in press.

19 Richard J. Herrnstein and Charles Murray, *The Bell Curve: Intelligence and Class Structure in American Life*, Free Press, New York, 1994.

2 APPLIED BIOLOGY

1 Leon Poliakov, *The Aryan Myth*, Chatto/Heinemann for Sussex University Press, London, 1974; John M. Efron, *Defenders of the Race: Jewish Doctors and Race Science in Fin-de-Siècle Europe*, Yale University Press, New Haven, 1994.

2 Poliakov, op.cit.

3 Nancy L. Stepan, *The Idea of Race in Science*, Macmillan, London, 1982.

4 Claude Blanckaert in *Bones, Bodies, Behavior: Essays on Biological Anthropology*, ed. George W. Stocking, University of Wisconsin Press, Madison, Wisconsin, 1988; Kenan Malik, *The Meaning of Race: Race, History and Culture in Western Society*, Macmillan, London, 1996.

5 Stephen Jay Gould, 'Petrus Camper's Angle', in *Bully for Brontosaurus*, Hutchinson Radius, London, 1991.

6 Nancy L. Stepan, 'Race and Gender: The Role of Analogy in Science', in *The 'Racial' Economy of Science: Towards A Democratic Future*, ed. Sandra Harding, Indiana University Press, Bloomington, 1993.

7 Elazar Barkan, *The Retreat of Scientific Racism: Changing Concepts of Race in Britain and the United States between the World Wars*, Cambridge University Press, Cambridge, 1992.

8 R.N. Bradley, *Racial Origins of English Character*, George Allen & Unwin, London, 1926.

9 Adolf Hitler, *Mein Kampf*, Hurst & Blackett, London, 1939.

10 Robert N. Proctor, *Racial Hygiene: Medicine under the Nazis*, Harvard University Press, Cambridge, Mass., 1988.

11 Ilse Schwidetzky, *History of Biological Anthropology in Germany*, International Association of Human Biologists, Newcastle, 1992; Proctor, op cit.

Notes and References

12 Michael Billig, 'Psychology, Racism and Fascism', A., F. & R. Publications, Birmingham, 1979.

13 Proctor in Stocking, 1988, op. cit.

14 Stefan Kühl, *The Nazi Connection: Eugenics, American Racism, and German National Socialism*, Oxford University Press, New York, 1994; Proctor, op. cit.

15 Schwidetzky, op. cit.

16 Proctor, op. cit. See also Schwidetzky, op. cit., for an example.

17 Schwidetzky, op. cit.

18 Ilse Schwidetzky, 'Neue Merkmalskarten von Mitteleuropa', *Zeitschrift für Rassenkunde* 14, 1943, 1–29.

19 Ilse Schwidetzky, *Zeitschrift für Rassenkunde* 13, 1942, 342 (review of *Weltkampf*); also reviews of *Forschungen zur Judenfrage* V, same page; *Forschungen zur Judenfrage* III, 9, 1939, 294–5; *Die Judenfrage in Ungarn*, 8, 1938, 215; *Judentum und Wissenschaft*, 6, 1937, 119.

20 Kühl, op. cit.

21 Proctor, op. cit.

22 Donna Haraway, *Primate Visions,* Verso, London, 1992.

23 Juan Comas, ' "Scientific" Racism Again?', *Current Anthropology* 2(4), 1961.

24 Ashley Montagu, *Statement on Race*, Oxford University Press, Oxford, 1972.

25 Jared Diamond, 'Race without Color', *Discover*, November 1994; William B. Provine, *Sewall Wright and Evolutionary Biology*, University of Chicago Press, Chicago, 1986.

26 Stephen Jay Gould, 'Human Equality is a Contingent Fact of History', *The Flamingo's Smile*, Penguin, Harmondsworth, Middlesex, 1986; Steven Pinker, *The Language Instinct*, Allen Lane, London, 1994.

27 Stocking, op. cit.

28 Ibid.

29 George W. Stocking Jr, *Race, Culture, and Evolution: Essays in the History of Anthropology*, University of Chicago Press, 1982.

30 William B. Provine, 'Geneticists and race', *American Zoologist* 26, 1986, 857–887.

31 Provine; Proctor in Stocking, 1988, op. cit.

32 Provine, op. cit.

33 William C. Boyd, *Genetics and the Races of Man*, Little, Brown, Boston, 1950; Carleton Coon, Stanley M. Garn and Joseph B. Birdsell, *Races: A Study of the Problems of Race Formation in Man*, Charles C. Thomas, Springfield, 1950; *Origin and Evolution of Man*, Cold Spring Harbor Symposium on Quantitative Biology 15, K.B. Warren, Philadelphia, 1950.

34 Proctor in Stocking, 1988, op. cit.

35 'Issues in the Study of Race: Two Views from Poland, with Discussion', *Current Anthropology* 3(1), February 1962.

Notes and References

36 T. Bielicki, T. Krupinski, J. Strzalko, *History of Physical Anthropology in Poland*, International Association of Human Biologists, Occasional Papers 1 (6), Newcastle upon Tyne, 1985.

37 Frank B. Livingstone, 'On the Non-Existence of Human Races', *Current Anthropology* 3(3), June 1962. For a recent popular discussion see Jared Diamond, *Discover*, November 1994.

38 See Sandra Harding, op. cit.

39 Alice Littlefield, Leonard Lieberman, Larry T. Reynolds, 'Redefining Race: The Potential Demise of a Concept in Physical Anthropology', *Current Anthropology* 23(6), 1982, 641–655.

40 Montagu, op. cit.

41 Ronald Walters, 'The Politics of the Federal Violence Initiative', Howard University (unpublished draft); Pat Shipman, *The Evolution of Racism*, Simon & Schuster, New York, 1994.

42 'Explanation of the Aims and Principles of the Northern League, a North-European Cultural Society', n.d.

43 Roger Pearson, *Eugenics and Race, Blood Groups and Race,* Folk and Race, London, 1966.

44 *Searchlight* 111, September 1984.

45 Ilse Schwidetzky, *Mankind Quarterly* 2(1), 1961, 10–12.

46 Bozo Škerlj, *Man* 60, 1960, 172–173.

47 Juan Comas, op. cit.; 'Racism Discussion Continues', *Current Anthropology* 3(3), June 1962.

48 Billig, op. cit.

49 *Mankind Quarterly* 15(3), Jan-Mar 1975.

50 Richard Lynn, 'Race Differences in Intelligence: A Global Perspective', *Mankind Quarterly* 1991, 31, 254–296.

51 Billig, op. cit.

52 *taz, die tageszeitung*, 3 September, 22 September, 9 October 1994.

53 Benno Müller-Hill, *Murderous Science: Elimination by Scientific Selection of Jews, Gypsies and Others, Germany 1933–1945*, Oxford University Press, Oxford, 1988.

54 Steve Jones, *The Language of the Genes: Biology, History and the Evolutionary Future*, Harper Collins, London, 1993.

55 Haraway, op. cit.

3 ANTIPODES

1 Jared Diamond, *The Rise and Fall of the Third Chimpanzee,* Hutchinson Radius, London, 1991.

2 John R. Baker, *Race,* Oxford University Press, Oxford, 1974.

3 Donna Haraway, *Primate Visions,* Verso, London, 1992.

4 Carleton Coon, *The Origin of Races,* Jonathan Cape, London, 1962.

5 Pat Shipman, 'On the Origin of Races', *New Scientist,* 16 January 1993.

6 Erik Trinkaus and Pat Shipman, *The Neandertals: Changing the Image of Mankind,* Jonathan Cape, London, 1993; Pat Shipman, *The Evolution of Racism,* Simon & Schuster, New York, 1994; Christopher Wills, *The Runaway Brain,* Harper Collins, London, 1994.

7 Theodosius Dobzhansky, *Current Anthropology* 4(4), October 1963.

8 Stephen Jay Gould, *Wonderful Life,* Hutchinson Radius, London, 1990.

9 See Robert W. Sussman, 'Overview', Contemporary Issues Forum: A Current Controversy in Human Evolution, *American Anthropologist* 95(1), March 1993.

10 R. L. Cann, M. Stoneking, A. C. Wilson, 'Mitochondrial DNA and Human Evolution', *Nature* 325, 1987, 31–36.

11 Alan Templeton, 'The "Eve" Hypothesis: A Genetic Critique and Reanalysis', Contemporary Issues Forum: A Current Controversy in Human Evolution, *American Anthropologist* 95(1), March 1993; interview with author.

12 C. L. Brace (book review), *Man* 29(2), June 1994, 474; Trinkaus and Shipman, op. cit.

13 Interview with author.

14 R.L. Kirk, *Aboriginal Man Adapting: The Human Biology of Australian Aborigines,* Clarendon, Oxford, 1981.

15 A. P. Elkin, *Current Anthropology* 2(4), October 1961, 317–8.

16 Robert Milliken, *No Conceivable Injury: The Story of Britain and Australia's Atomic Cover-up,* Penguin Books Australia, Ringwood, Victoria, 1986.

17 *Cambridge Encyclopedia of Human Evolution,* eds Steve Jones, Robert Martin and David Pilbeam, Cambridge University Press, Cambridge, 1992.

18 G. Richard Scott and Christy G. Turner II, 'Dental Anthropology', *Annual Review of Anthropology 1988* 17, 99–126.

19 C. Loring Brace, 'Australian Tooth Size Clines and the Death of a Stereotype', *Current Anthropology* 21, 1980, 141–164.

20 Claude Lévi-Strauss, *Race and History,* Unesco, Paris, 1968.

21 Haraway, op. cit.

22 Carleton Coon, *The Hunting Peoples,* Jonathan Cape, London, 1972.

4 CAN WHITE MEN JUMP?

1 John H. Himes, 'Racial Variation in Physique and Body Composition', *Canadian Journal of Sport Sciences* 13(2), 1988, 117–126.

2 *Observer, Life* magazine, 22 January 1995.

3 Jonathan Kingdon, *Self-Made Man and His Undoing,* Simon & Schuster, London, 1993.

4 Vincent Sarich, University of California at Berkeley, Anthropology 1 lecture notes, transcription by Late Night Notes, Berkeley, California, 1990.

5 Amby Burfoot, 'White Men Can't Run', in *The Best American Sports Writing 1993*, ed. Frank Deford, Houghton Mifflin, Boston, 1993. The article originally appeared in *Runner's World*, August 1992.

6 Interview with author, April 1995.

7 *Running Research News*, March–April 1993.

8 Fatimah Linda Collier Jackson, 'Evolutionary and Political Economic Influences on Biological Diversity in African Americans', *Journal of Black Studies* 23(4), June 1993, 539–560.

9 Jacques Samson and Magdaleine Yerlès, 'Racial Differences in Sports Performances', *Canadian Journal of Sport Sciences* 13(2), 1988, 109–116.

10 *Sun*, 17 August 1993.

11 Matthew 25: 14–30 (Douay Version).

12 Claude Bouchard, 'Brief Review: Racial Differences in Performance', *Canadian Journal of Sport Sciences* 13(2), 1988, 103.

13 Samson and Yerlès, op. cit.

14 Robert M. Malina, 'Racial/Ethnic Variation in the Motor Development and Performance of American Children', *Canadian Journal of Sport Sciences* 13(2), 1988, 136–143.

15 Himes, op. cit.

16 *Cambridge Encyclopedia of Human Evolution*, eds Steve Jones, Robert Martin and David Pilbeam, Cambridge University Press, Cambridge, 1992.

17 See Dave Hill, *Out Of His Skin: The John Barnes Phenomenon*, Faber and Faber, London, 1989.

18 Robert Malina, quoted by Samson and Yerlès, op. cit.

19 Marcel R. Boulay, Pierre F. M. Ama and Claude Bouchard, 'Racial Variation in Work Capacities and Powers', *Canadian Journal of Sport Sciences* 13(2), 1988, 127–135.

5 FIFTEEN POINTS

1 Charles Locurto, 'On the Malleability of IQ', *The Psychologist: Bulletin of the British Psychological Society* 11, 1988, 431–435; Richard P. Feynman, '*Surely You're Joking, Mr. Feynman!' Adventures of a Curious Character*, Norton, New York, 1985.

2 Steven Rose, R. C. Lewontin and Leon J. Kamin, *Not in Our Genes: Biology, Ideology and Human Nature*, Penguin, London, 1990.

3 Stephen Jay Gould, *The Mismeasure of Man*, W.W. Norton, New York, 1981.

4 See Peter Schönemann, 'Jensen's *g*: Outmoded Theories and Unconquered

Notes and References

Frontiers', in *Arthur Jensen, Consensus and Controversy*, eds Sohan Modgil and Celia Modgil, Falmer Press, New York, 1987.

5 Mark Snyderman, 'How to Think about Race' (book review), *National Review*, 12 September 1994.

6 Richard J. Herrnstein and Charles Murray, *The Bell Curve: Intelligence and Class Structure in American Life*, Free Press, New York, 1994.

7 Daniel Seligman, *A Question of Intelligence: The IQ Debate in America*, Citadel Press, New York, 1994.

8 Stephen Jay Gould, 'Jensen's Last Stand', *An Urchin in the Storm*, Collins Harvill, London, 1988.

9 Seligman, op. cit.

10 Modgil and Modgil, op. cit.

11 Stephen Jay Gould, 'Jensen's Last Stand', op. cit.

12 Herrnstein and Murray, op. cit.

13 See Steven Pinker, *The Language Instinct*, Allen Lane, London, 1994.

14 Letter to the *New Yorker*, 26 December 1994/2 January 1995.

15 Seligman, op. cit.

16 Ibid.

17 Christopher Brand, *Nature* 368, 10 March 1994, 111.

18 Charles Murray and Richard J. Herrnstein, 'Race, Genes and I.Q. – An Apologia', *New Republic*, 31 October 1994.

19 Seligman, op. cit.

20 Modgil and Modgil, op. cit.

21 Cornel West, *Race Matters*, Beacon, Boston, Mass., 1993.

22 Ian J. Deary, 'Pandora's Box and the Eskimo's Nose', *Mankind Quarterly* 32(1–2), 1991, 153–159.

23 Thomas Bouchard *et al.*, 'Sources of Human Psychological Differences: The Minnesota Study of Twins Reared Apart', *Science* 250, 1990, 223–228.

24 Seligman, op. cit.; James R. Flynn, 'Japanese Intelligence Simply Fades Away: A Rejoinder to Lynn (1987)', *The Psychologist: Bulletin of the British Psychological Society* 9, 348–350, 1988; 'Massive IQ Gains in 14 Nations: What IQ Tests Really Measure', *Psychological Bulletin* 101(2), 1987, 171–191.

25 Modgil and Modgil, op. cit.

26 Juan Comas, *Current Anthropology* 2(4), 1961, 335; 3(3), 1962, 302.

27 Herrnstein and Murray, op. cit.; Sandra Scarr, 'Three Cheers for Behavior Genetics: Winning the War and Losing Our Identity', *Behavior Genetics* 17 (3), 1987, 219–228.

28 *New York Times* magazine, 9 October 1994.

29 Sandra Scarr, 'Human Differences and Political Equality: The Dilemma of

Group Differences in IQ' (review of *The Bell Curve*), *Issues in Science and Technology* XI (2), Winter 1994/1995, 82–85.

30 Scarr, 1987, op. cit.

31 A. M. West, N. J. Mackintosh, C. G. N. Mascie-Taylor, 'Cognitive and Educational Attainment in Different Ethnic Groups', *Journal of Biosocial Science* 24, 1992, 539–554. See also N. J. Mackintosh, 'The Biology of Intelligence?', *British Journal of Psychology* 77, 1986, 1–18.

32 James R. Flynn, *Race, IQ and Jensen*, Routledge & Kegan Paul, London, 1980.

33 Herrnstein and Murray, op. cit.

34 Modgil and Modgil, op. cit.

35 Adrian Wooldridge, *Measuring the Mind: Education and Psychology in England, c.1860–1990*, Cambridge University Press, Cambridge, 1994.

36 'America's Wasted Blacks', *The Economist*, 30 March 1991.

37 Mark Snyderman and Stanley Rothman, *The IQ Controversy: The Media and Public Policy*, Transaction, New Brunswick, N.J., 1988.

38 Seligman, op. cit.

39 These comments are based on an interview with J. P. Rushton, February 1994.

40 Charles Lane, *New York Review of Books*, 1 December 1994.

41 Barry Mehler, 'Foundation for Fascism: The New Eugenic Movement in the United States', *Patterns of Prejudice* 23(4), 1989, 17–25; Stefan Kühl, *The Nazi Connection: Eugenics, American Racism, and German National Socialism*, Oxford University Press, New York, 1994.

42 *Guardian*, 18 July 1992.

43 Adam Miller, 'Professors of Hate', *Rolling Stone*, 20 October 1994.

44 Kühl, op. cit.

45 Miller, op. cit. (excludes 1983 grants).

46 As in 'Rolonda', King World to NBC, 1994.

47 Mehler, op. cit.; *Guardian*, 18 July 1992; *Independent on Sunday*, 4 March 1990; *Sunday Telegraph*, 12 March 1989; *Independent*, 8 March 1989.

48 *Wall Street Journal Europe*, 10 January 1995.

49 *Science* 266, 16 December 1994, 1811.

50 Raymond B. Cattell, *Beyondism: Religion from Science,* Praeger, New York, 1987. Cf. Robert Shapiro (*The Human Blueprint*, St Martin's Press, New York, 1991), who also anticipates biological divergence, into subspecies rather than species, as a result of genetic engineering.

51 Interview with author.

52 Stephen Jay Gould, 'Nurturing Nature', *An Urchin in the Storm*, Collins Harvill, London, 1988; *International Herald Tribune*, 30 December 1993; 'China's Misconception of Eugenics', *Nature* 367, 6 January 1994, 1, 3.

53 See Studs Terkel, *Race*, Sinclair-Stevenson, London, 1992.

54 Paul M. Sniderman and Thomas Piazza, *The Scar of Race*, Belknap, Cambridge, Mass., 1993.

55 National Opinion Polls for the Runnymede Trust and the *Independent on Sunday*; see Kaushika Amin and Robin Richardson, *Politics for All: Equality, Culture and the General Election 1992*, Runnymede Trust, London, 1992; *Independent on Sunday*, 7 July 1991.

56 *New Republic*, 31 October 1994.

57 Stephen Jay Gould, 'Nurturing Nature', *An Urchin in the Storm*, Collins Harvill, London, 1988.

58 Ibid.

59 *New York Review of Books*, 17 November 1994.

6 THE HIGHER LATITUDES

1 R. Lynn, 'Oriental Americans: Their IQ, Educational Attainment and Socio-Economic Status' (Special Review), *Personality and Individual Differences* 15(2), 1993, 237–242.

2 Ibid.

3 Richard Lynn, 'Reply to Commentaries on Racial Differences in Intelligence', *Mankind Quarterly* 32 (1–2), 1991, 161–173.

4 *Sunday Telegraph*, 12 June 1994.

5 Interview with author; Richard Lynn, 'Orientals: An Emerging American Elite?' *Mankind Quarterly* 31 (1–2), 1990, 185–190.

6 Letters, *Spectator*, 14 January 1995.

7 Lynn, 1990, op. cit.

8 Richard Lynn, James R. Flynn, Ch. 20, *Intelligence: Ethnicity and Culture, Cultural Diversity and the Schools*, eds J. Lynch, C. Modgil, S. Modgil, Falmer, London, 1992; Richard Lynn, 1993, op. cit.

9 Richard Lynn, 'Race Differences in Intelligence: A Global Perspective', *Mankind Quarterly* 31, 1991, 254–296.

10 James R. Flynn, *Race, IQ and Jensen*, Routledge & Kegan Paul, London 1980.

11 Roger Omond, *The Apartheid Handbook*, Penguin, Harmondsworth, Middlesex, 1985.

12 Charles Lane, 'The Tainted Sources of "The Bell Curve" ', *New York Review of Books*, 1 December 1994. See also Leon J. Kamin, *Scientific American*, February 1995.

13 Richard Lynn, 'The Intelligence of the Mongoloids: A Psychometric, Evolutionary and Neurological Theory', *Personality and Individual Differences* 8(6), 1987, 813–844.

14 Joy Hendry, *Becoming Japanese: The World of the Pre-school Child*, Manchester University Press, Manchester, 1986.

15 Harold W. Stevenson, 'Learning from Asian Schools', *Scientific American*, December 1992, 328–8.

16 'Wot U Lookin At?', written and produced by Oliver James and David Malone, *Horizon*, BBC, 1993.

17 Richard Lynn, *The Secret of the Miracle Economy, Different National Attitudes to Competitiveness and Money*, Social Affairs Unit, London, 1991.

18 Richard Lynn, 'The Evolution of Racial Differences in Intelligence', *Mankind Quarterly* 32, (1–2), 1991, 99–121.

19 Ibid.

20 Rhys Jones, 'East of Wallace's Line: Issues and Problems in the Colonisation of the Australian Continent', in *The Human Revolution: Behavioural and Biological Perspectives on the Origins of Modern Humans*, eds Paul Mellars and Christopher Stringer, Edinburgh University Press, Edinburgh, 1989.

21 R. Lynn, 'Race Differences in Intelligence: A Global Perspective,' *Mankind Quarterly* 31, 1991, 254–296; J. Klekamp et al., 'Morphometric Study on the Postnatal Growth of the Visual Cortex of Australian Aborigines and Caucasians', *Journal of Brain Research* 35 (4), 1994, 541–8.

22 See Robin I. M. Dunbar in *The Sociobiology of Ethnocentrism*, eds. V. Reynolds, Vincent S. E. Falger, Ian Vine, Croom Helm, London, 1987.

23 Christopher Stringer and Clive Gamble, *In Search of the Neanderthals*, Thames & Hudson, London, 1993.

24 Maria T. Phelps, 'An Examination of Lynn's Evolutionary Account of Racial Differences in Intelligence', *Mankind Quarterly* 33(3), 1993, 295–308.

25 Interview with author.

26 Ashley Montagu, *Current Anthropology* 2(4) 1961, 323–326.

27 Peter Høeg, *Miss Smilla's Feeling for Snow*, Flamingo, London, 1994.

28 Philip E. Vernon, *Intelligence: Heredity and Environment*, W. H. Freeman, San Francisco, 1979.

29 Ian Creery, *The Inuit (Eskimo) of Canada*, MRG International Report 93 /3, Minority Rights Group, London, 1993.

30 *History and Geography of Human Genes*, eds L. Luca-Cavalli Sforza, Paolo Menozzi and Alberto Piazza, Princeton University Press, Princeton, New Jersey, 1994.

31 R. Lynn, 'The Evolution of Racial Differences in Intelligence', *Mankind Quarterly* 32 (1–2), 1991, 99–121.

32 *Cambridge Encyclopedia of Human Evolution*, eds Steve Jones, Robert Martin and David Pilbeam, Cambridge University Press, Cambridge, 1992.

33 Geoffrey Miller, 'The History of Passion: A Review of Sexual Selection and Human Evolution', in *Evolution and Human Behavior: Ideas, Issues, and Applications*, ed. C. Crawford, Lawrence Erlbaum, Hillsdale, N.J., in press; Matt

Notes and References

Ridley, *The Red Queen: Sex and the Evolution of Human Nature*, Viking, London, 1993.

34 Christopher Wills, *The Runaway Brain*, Harper Collins, London, 1994.

35 Phillip V. Tobias, *The Brain in Hominid Evolution*, Columbia University Press, New York, 1971.

36 M. Henneberg et al., 'Head Size, Body Size and Intelligence: Intraspecific Correlations in *Homo sapiens sapiens*', *Homo* 36, 1985, 207–218; Roger Lewin, 'Rise and Fall of Big People', *New Scientist*, 22 April 1995.

37 Stephen Jay Gould, *The Mismeasure of Man*, W. W. Norton, New York, 1981.

38 R. Lynn, 'The Evolution of Racial Differences in Intelligence'; 'Further Evidence for the Existence of Race and Sex Differences in Cranial Capacity', *Social Behavior and Personality* 21(2), 1993, 89–92; J. Philippe Rushton, 'Mongoloid-Caucasoid Differences in Brain Size from Military Samples', *Intelligence* 15, 1991, 351–359.

39 Kenneth L. Beals, Courtland L. Smith, Stephen M. Dodd, 'Brain Size, Cranial Morphology, Climate and Time Machines', *Current Anthropology* 25(3), 1984, 301–330.

40 Stephen Jay Gould, 'Sizing up Human Intelligence', *Ever Since Darwin: Reflections in Natural History*, Penguin, London, 1980.

41 See L. M. Van Valen, *Nature* 359, 29 October 1992, 768 (correspondence).

42 M. Henneberg et al., op. cit.

43 Kenneth L. Beals, 'Problems and Issues with Human Brain Size, Body Size and Cognition', *Homo* 37(3), 1986, 148–60.

44 J. P. Rushton, *Race, Evolution, and Behavior: A Life History Perspective*, Transaction, New Brunswick, N.J., 1994.

45 Ibid.; N. Raz et al., 'Neuroanatomical Correlates of Age-sensitive and Age-invariant Cognitive Abilities: An *in vivo* MRI Investigation', *Intelligence* 17, 1993, 407–422.

46 John C. Wickett, Philip A. Vernon, Donald H. Lee, '*In vivo* Brain size, Head perimeter and Intelligence in a Sample of Healthy Adult Females', *Personality and Individual Differences* 16(6), 1994, 831–8.

47 J. P. Rushton, 'Mongoloid-Caucasoid Differences in Brain Size from Military Samples', *Intelligence* 15, 1991, 351–359; 'Cranial Capacity Related to Sex, Rank and Race in a Stratified Random Sample of 6, 325 U.S. Military Personnel', *Intelligence* 16, 1992, 401–413.

48 Lee Willerman, 'Commentary on Rushton's Mongoloid-Caucasoid Differences in Brain Size', *Intelligence* 15, 1991, 361–364; and J. P. Rushton's 'Reply to Willerman on Mongoloid-Caucasoid Differences in Brain Size,' 365–7; T. Edward Reed and Arthur R. Jensen, 'Cranial Capacity: New Caucasian Data

297

Notes and References

and Comments on Rushton's Claimed Mongoloid-Caucasoid Brain-size Differences', *Intelligence* 17, 1993, 423–431.

49 J. P. Rushton, 'Sex and Race Differences in Cranial Capacity from International Labour Office Data', *Intelligence* 19, 1994, 281–294.

50 John Maddox, 'How to Publish the Unpalatable?' (editorial), *Nature* 358, 16 July 1992, 187; correspondence: *Nature* 358, 13 August 1992, 532; 359, 17 September 1992, 181–182; 359, 29 October 1992, 768; 360, 26 November 1992, 292.

51 C. Brand, *Nature* 359, 29 October 1992, 768.

52 Richard Lynn, 'Sex Differences in Intelligence and Brain Size: A Paradox Resolved', *Personality and Individual Differences* 17(2), 1994, 257–272.

53 J. P. Rushton, *Race*, 1994.

54 J. P. Rushton, 'Race Differences in Behaviour: A Review and Evolutionary Analysis', *Personality and Individual Differences* 9(6), 1988, 1009–1024.

55 J. P. Rushton, *Race*, 1994.

56 Adam Miller, 'Professors of Hate', *Rolling Stone*, 20 October 1994.

57 A French Army Surgeon, *Untrodden Fields of Anthropology: Observations on the Esoteric Manners and Customs Of Semi-Civilised Peoples, Being a Record by a French Army Surgeon of Thirty Years' Experience in Asia, Africa, and America*, Charles Carrington, Paris, 1896.

58 J. P. Rushton, 'Do r-K Strategies Underlie Human Race Differences? A reply to Weizmann *et al.*', *Canadian Psychology* 32(1), 1991, 29–42.

59 J. M. Diamond, 'Variation in Human Testis Size', *Nature* 320, 10 April 1986, 488–489.

60 P. H. Harvey and R. M. May, 'Out for the Sperm Count', *Nature* 337, 9 February 1989, 508–509.

61 Richard J. Herrnstein and Charles Murray, *The Bell Curve: Intelligence and Class Structure in American Life*, Free Press, New York, 1994.

62 G. Allen et al., 'Twinning and the r/K Reproductive Strategy: A Critique of Rushton's Theory', *Acta Geneticae, Medicae, et Gemellologiae* 41(1), 1992, 73–83; Vernon Reynolds in *The Sociobiological Imagination*, ed. Mary Maxwell, State University of New York Press, Albany, 1991.

63 Kristen Hawkes, J. P. Rushton, James S. Chisholm, 'On Life-History Evolution', *Current Anthropology* 35(1), 1994, 39–46.

64 Donald C. Johanson and Maitland A. Edey, *Lucy: The Beginnings of Humankind*, Penguin, London, 1990; J. P. Rushton, 'An Evolutionary Theory of Health, Longevity and Personality: Sociobiology and r/K Reproductive Strategies', *Psychological Reports* 60, 1987, 539–549; Contributions to the history of psychology: XC. Evolutionary biology and heritable traits (with reference to Oriental-black-white differences): The 1989 AAAS Paper, *Psychological Reports* 71, 1992, 811–821.

Notes and References

65 J. P. Rushton, 'The Evolution of Racial Differences: A Reply to M. Lynn', *Journal of Research in Personality*, 1989, 23, 7–20.

66 See for example Robert Wright, *The Moral Animal*, Little, Brown, London, 1995.

67 Marvin Zuckerman and Nathan Brody, 'Oysters, Rabbits and People: A Critique of "Race Differences in Behaviour" by J. P. Rushton', *Personality and Individual Differences* 9(6), 1988, 1025–1033; David Barash, review of J. P. Rushton, *Race*, 1994, *Animal Behaviour* 49 (4), 1995, 1131–3.

68 J. P. Rushton, *Race*, 1994, 'Race and Crime: An International Dilemma', *Society* 32, 1995, 37–41.

69 Interview with author.

70 Richard Leakey and Roger Lewin, *Origins Reconsidered: In Search of What Makes Us Human*, Little, Brown, London, 1992.

71 Interview with author.

72 Phyllis B. Eveleth and James M. Tanner, *Worldwide Variation in Growth and Development*, Cambridge University Press, Cambridge, 1990.

73 J. P. Rushton, 1988, op. cit.

74 Interview with author.

75 Roger Pearson, *Race, Intelligence and Bias in Academe*, Scott-Townsend, Washington D.C., 1991; *Rolling Stone*, 20 October 1994.

76 J. P. Rushton, *Race*, 1994.

77 J. P. Rushton, *Race*, 1994; Lee Ellis, 'Criminal Behaviour and r/K selection: An Extension of Gene-based Evolutionary Theory', *Personality and Individual Differences* 9(4), 1988, 697–708; also published in *Deviant Behavior* 8, 1987, 149–176.

78 Michael Lynn, 'Criticism of an Evolutionary Hypothesis about Race Differences: A Rebuttal to Rushton's Reply', *Journal of Research in Personality*, 1989, 23, 21–34; Irwin Silverman, 'The r/K Theory of Human Individual Differences: Scientific and Social Issues', *Ethology and Sociobiology* 11, 1990, 1–9.

79 Interview with author.

80 J. P. Rushton, *Race*, 1994.

81 Edward M. Miller, 'Could r Selection Account for the African Personality and Life Cycle?', *Personality and Individual Differences* 15(6), 1993, 665–675; J. P. Rushton, *Race*, 1994.

82 F. L. C. Jackson, 'Evolutionary and Political Economic Influences on Biological Diversity in African Americans', *Journal of Black Studies* 23(4), 1993, 539–560.

83 F. L. C. Jackson, 'Race and Ethnicity as Biological Constructs', *Ethnicity and Disease* 2, 1992, 120–125; 'The Bioanthropological Context of Disease', *American Journal of Kidney Diseases* 21(4, Suppl. 1), 1993, 10–14.

84 J. P. Rushton, *Race*, 1994.

85 Interview with author. In full, the relevant exchange runs as follows:

– Do you consider yourself to be a racialist, in the sociopolitical sense of the term?

'I think the primary meaning of a racialist is one who believes there are race differences in various characteristics, particularly things like intelligence and crime. In that sense I am, yes.

'A secondary sense is that we should therefore persecute races, which I don't subscribe to.'

– Do you differentiate between the terms 'racialist' and 'racist'?

'Not really, no.'

– So you are a racist in the sense you just described?

'Sometimes it's called a scientific racist, isn't it? In that sense I am, yes.'

86 Interview with author:

'I don't consider myself a racist.'

– Not even a scientific racist?

'Not at all. A racist to me is somebody who categorises people all the same, in a group, and I've already stated there's enormous overlap, so you can't judge individuals that way, and racists typically want to classify people into a group, typologically, in order to mistreat them. I don't want to mistreat anybody.'

87 A. Miller, op. cit.

88 *The Times*, 24 October 1994.

89 *Nature* 347, 6 September 1990, 6; Barry R. Gross, 'The Case of Philippe Rushton', *Academic Questions* 3(4), 1990, 35–46; Roger Pearson, *Race, Intelligence and Bias in Academe*, Scott-Townsend, Washington D.C., 1991.

90 Hans J. Eysenck, *Rebel with a Cause*, W. H. Allen, London, 1990.

91 *Sunday Telegraph*, 14 August 1994.

92 *Sunday Telegraph*, 11, 25 September, 2 October 1994.

7 CAVE MEN WITH ATTITUDE

1 Richard Lynn, 'Race Differences In Intelligence: A Global Perspective', *Mankind Quarterly* 31, 1991, 254–296.

2 John R. Baker, *Race*, Oxford University Press, Oxford, 1974.

3 *Beacon* 1, 1977: 'Fortunately, we have recently had a very interesting book by Prof. Baker called *Race* and he has come to the conclusion that there is a very close correlation between the different achievements of races and their present day IQ level.'

4 Cheikh Anta Diop, *Civilisation or Barbarism: An Authentic Anthropology*, Lawrence Hill Books, Brooklyn, New York, 1991.

5 Kwame Anthony Appiah, *The Times Literary Supplement*, 12 February 1993.

Notes and References

6 Molefi Kete Asante, *Afrocentricity: The Theory of Social Change*, Amulefi, Buffalo, New York, 1980.

7 Appiah, op. cit.

8 Martin Bernal, *Black Athena*, Vintage, London, 1991.

9 Bernard Ortiz de Montellano, 'Melanin, Afrocentricity, and Pseudoscience', *Yearbook of Physical Anthropology*, 36, 1993, 33–58; C. Loring Brace *et al.*, 'Clines and Clusters Versus "Race": A Test in Ancient Egypt and the Case of a Death on the Nile', *Yearbook of Physical Anthropology* 36, 1993, 1–31.

10 Diop, op. cit.

11 *Science* 262, 12 November 1993, 1121–1122.

12 Ivan Van Sertima (ed.), *Blacks in Science, Ancient and Modern*, Transaction, New Brunswick, New Jersey, 1983; Paul R. Gross and Norman Levitt, *Higher Superstition: The Academic Left and its Quarrels with Science*, Johns Hopkins University Press, Baltimore, Maryland, 1994.

13 Bernard Ortiz de Montellano, 'Multicultural Pseudoscience: Spreading Scientific Illiteracy Among Minorities – Part 1', *Skeptical Inquirer* 16(1), 1991, 46–50.

14 'Bones of Contention', written and produced by Danielle Peck and Alex Seaborne, *Horizon*, BBC, 1995.

15 *Guardian*, 24 February 1994.

16 Gloria Thomas-Emeagwali (ed.), *African Systems of Science, Technology and Art: The Nigerian Experience*, Karnak House, London, 1993; cf. Diop, op. cit.

17 Yehudi Webster, 'Afrocentrism, Racism, and Other Myths' (audio tape), Skeptics Society, Altadena, California, 1994.

18 Stephen Howe, unpublished manuscript; Ortiz de Montellano, 1993, op. cit.

19 Bernard Ortiz de Montellano, 'Magic Melanin: Spreading Scientific Illiteracy Among Minorities – Part 2', *Skeptical Inquirer* 16(2), 1992, 162–166; 1993, op. cit.

20 http://www.melanet.com/melanet/ubus/melib.html

21 *Sunday Times*, 3 April 1994.

22 Stephen Howe, unpublished manuscript; Ortiz de Montellano, 1993, op. cit.

23 Ortiz de Montellano, 1992, op. cit.; Isaac Asimov, *Asimov's Biographical Encyclopedia of Science and Technology*, David & Charles, Newton Abbot, 1978.

24 Gross and Levitt, op. cit; Robert Hughes, *The Culture of Complaint*, Oxford University Press, New York, 1993

25 Cf. Appiah, op. cit.

26 Webster, op. cit.; 'The Hearts and Minds of City College', *New Yorker*, 7 June 1993.

27 *Independent*, 9 August 1993.

28 *Observer, Life* magazine, 6 November 1994 (interviewed by author).

29 Michael Bradley, *The Iceman Inheritance: Prehistoric Sources of Western Man's Racism, Sexism and Aggression*, Kayode, New York, 1991.

30 Diop, op. cit.

31 Hughes, op. cit.

32 *Guardian*, 1 April 1994.

33 Ice Cube, *Lethal Injection*, 4th & Broadway Records, London, 1993.

34 Christopher Wills, *Discover*, November 1994; Ortiz de Montellano, 1993, op. cit.

35 *Village Voice*, 21 December 1993.

36 'Black Demagogues and Pseudo-Scholars', *New York Times*, 20 July 1992.

37 *New Republic*, 31 October 1994.

8 TRIBAL LAW

1 Douglas Johnson, 'The New Right in France', in *Neo-fascism in Europe*, eds Luciano Cheles, Ronnie Ferguson, Michalina Vaughan, Longman, London, 1991; Ian R. Barnes in *The Dark Side of Europe: The Extreme Right Today*, ed. Geoffrey Harris, Edinburgh University Press, Edinburgh, 1990; Ian R. Barnes, 'Pedigree of GRECE', *Patterns of Prejudice* 14(3), 1980.

2 Michael Billig, *Psychology, Racism and Fascism*, A. F. and R. Publications, Birmingham, 1979.

3 *Searchlight*, April 1995; *Guardian*, 25 November 1994.

4 David Barash, *Sociobiology: The Whisperings Within*, Souvenir, London, 1980.

5 See Irwin Silverman in *The Sociobiology of Ethnocentrism: Evolutionary Dimensions of Xenophobia, Discrimination, Racism and Nationalism*, eds Vernon Reynolds, Vincent Falger, Ian Vine, Croom Helm, London, 1987.

6 Interview with author, March 1994.

7 Haraway, *Primate Visions*, Verso, London, 1992.

8 Pierre L. van den Berghe, 'Ethnicity and the Sociobiology Debate', in *Theories of Race and Ethnic Relations*, eds John Rex and David Mason, Cambridge University Press, Cambridge, 1986; 'Does Race Matter?', *Ethnic and Racial Studies*, in press.

9 M. Vaughan, 'The Extreme Right in France: 'Lepénisme' or the Politics of Fear', Cheles et al., op. cit.

10 Irenäus Eibl-Eibesfeldt, *The Biology of Peace and War: Men, Animals and Aggression*, Thames & Hudson, London, 1979.

11 Umberto Melotti, in Vernon Reynolds et al., op. cit.; J. P. Rushton, 'Genetic Similarity, Human Altruism, and Group Selection', *Behavioral and Brain Sciences* 12(3), 1989, 503–559; J. P. Rushton, *Altruism, Socialization, and Society*, Prentice-Hall, Englewood Cliffs, N.J., 1980; Niles Eldredge, *Reinventing Darwin: The Great Debate at the High Table of Evolutionary Theory*, John Wiley, New York, 1995.

12 Pierre L. van den Berghe, 'Does Race Matter?', op. cit.

13 Ibid.

14 Alex de Waal, 'Genocide in Rwanda', *Anthropology Today* 10(3), June 1994, 1–2.

15 *Asen Bulletin*, 5, September 1993.

16 P. L. van den Berghe, *Race and Racism, A Comparative Perspective*, John Wiley, New York, 1978.

17 V. Sarich/Late Night Notes, Berkeley, California, 1990.

18 Adam Miller, 'Professors of Hate', *Rolling Stone*, 20 October 1994.

19 *Science* 251, 1991, 368–371.

20 Robert Hughes, *The Culture of Complaint*, Oxford University Press, New York, 1993.

21 V. Sarich, 'Race and Language in Prehistory', 1995, in press.

22 Sarich, op. cit., 1990; 'In Defense of the Bell Curve: The Reality of Race and Human Differences', *Skeptic* 3(3), 1995, 84–94; pers. comm.

9 SOCIALLY UNADAPTABLE

1 AP 6 September 1993, BBC Monitoring Service, Summary of World Broadcasts, RTN 8 September 1993.

2 Benno Müller-Hill, *Murderous Science: Elimination by Scientific Selection of Jews, Gypsies and Others, Germany 1933–1945*, Oxford University Press, Oxford, 1988.

3 BBC, SWB 15 September 1993 Slovakia: TV Polonia satellite service, Warsaw, in Polish 1900 gmt, 9 September 1993.

4 BBC SWB, RTN 8 September 1993, RTN 15 September 1993.

5 Benno Müller-Hill, op. cit..

6 Ian Hancock in *The Gypsies of Eastern Europe*, eds David Crowe and John Kolsti, M. E. Sharpe, Armonk, New York, 1991.

7 Paul Hockenos, *Free to Hate: the Rise of the Ultra-Right in Eastern Europe*, Routledge, London, 1993.

8 Crowe and Kolsti, op. cit.

9 M. Kvasnicová et al., 'Geneticky Podmienena Mentalna Retardacia v Okrese Banská Bystrica', *Československa Pediatrie* 47(1), 1992, 25–8.

10 *Prague Post*, 16–22 December 1992.

11 M. Lescisinová et al., 'Increased Incidence of Congenital Hypothyroidism in Gypsies in East Slovakia as compared with White Population', *Endocrinologia Experimentalis* 23(2), 1989, 137–41.

12 J.D. Wilson and D.W. Foster, *Textbook of Endocrinology*, W.B. Saunders, Philadelphia, 1985.

13 *European*, 27–30 September 1991; Penn & Schoen Associates for Freedom House and American Jewish Committee, April 1991; *Prognosis*, 4 February 1994.

14 *San Francisco Examiner*, 19 December 1993; *Independent*, 19 October 1993. These two accounts differ in detail.

15 Ian Hancock, 'Anti-Gypsyism in the New Europe', *Roma* 38/39, 1993, 5–29.

16 Ibid.; *New York Times* Magazine, 24 June 1990.

17 Cited by Stephen Jay Gould, *The Mismeasure of Man*, W.W. Norton, New York, 1981.

18 Donald Kenrick, 'The situation of Gypsies in Romania', Association of Gypsy Organisations, London, September 1993.

19 Katherine Verdery, 'Nationalism in Romania', *Slavic Review* 52(2), 1993, 179–203.

20 Thomas S. Szayna, 'Ultra-Nationalism in Central Europe', *Orbis* 37(4), 1993, 527–550.

21 'Setting the Record Straight', *Magyar Forum* 20, August 1992; Judith Pataki, *RFE/RL Research Reports* 40, 9 October 1992, 15–12. Hockenos gives a different translation.

22 Edith Oltay, *RFE/RL Research Reports* 2(13), 26 March 1993, 26–29.

23 Liz Fekete and Frances Webber, *Inside Racist Europe*, Institute of Race Relations, London, 1994.

24 Chris Powell, 'Time for another Immoral Panic? The case of the Czechoslovak Gypsies', *International Journal of Sociology and Law* 2, 1994, 105–121.

25 Luba Kalaydjieva, Ivo Kremensky, 'Screening for Phenylketonuria in a Totalitarian State', *Journal of Medical Genetics* 29, 1992, 656–658.

26 D. Kadanov, S. Mutafov, *Cherepat na chovek v mediko-antropologichen aspekt (The human skull in a medico-anthropological aspect)*, Bulgarian Academy of Science, Sofia, 1984.

27 Carol Silverman in *Nomadic Peoples*, no. 21/22, December 1986, *Peripatetic Peoples*, eds Joseph C. Berland, Matt T. Salo, Commission on Nomadic Peoples, Montreal.

28 *Courrier International* No. 209, 3–9 November 1994.

29 Ian Hancock in Crowe and Kolsti, op. cit.; *Struggling for Identity: Czechoslovakia's Endangered Gypsies*, Helsinki Watch, Human Rights Watch, New York, 1992. See also J.A. Valšík, *Current Anthropology* 3(3), 1962, 294–298.

30 Chris Powell, op. cit.

31 Ibid.

32 T. Bielicki, T. Krupinski, J. Strzalko, *History of Physical Anthropology in Poland*, International Association of Human Biologists, Occasional Papers 1(6), Newcastle upon Tyne, 1985.

33 Leon Poliakov, *The Aryan Myth*, Chatto/Heinemann for Sussex University Press, London, 1974.

34 Ivan Bernasovský, *Seroanthropology of Roms (Gypsies)*, P. J. Šafárik University, Košice, 1994; Interview with author.

35 K. Bernasovská et al., 'Proposal of Low Birth Weight Limit for Gypsy Mature Babies', *Anthropology of Maternity*, Charles University, Prague, 173–175.

36 K. Hughes, N. R. Tan, K. C. Lun, 'Ethnic Group Differences in Low

Birthweight of Live Singletons in Singapore, 1981–3', *Journal of Epidemiology and Community Health*, 40(3), 1986, 262–6; I.R. McFadyen et al., 'Factors Affecting Birthweights in Hindus, Moslems and Europeans', *British Journal of Obstetrics and Gynaecology* 91(10), 1984, 968–72.

37 M. I. Levene, D. Tudehope, J. Thearle, *Essentials of Neonatal Medicine*, Blackwell Scientific Publications, Oxford, 1987; M. Wilcox et al., 'Birth Weight from Pregnancies Dated by Ultrasonography in a Multicultural British Population', *British Medical Journal*, 4 September 1993, 307(6904), 588–591.

38 K. Joubert, 'Correlation between Somatic Developmental Status of Newborn Infants and Some Sociodemographic Data', Central Statistical Office, Demographic Research Institute, Budapest, *Arztliche Jugendkunde* 81(5), 1990, 347–52.

39 P. L. Yudkin, S. Harlap, M. Baras, 'High Birthweight in an Ethnic Group of Low Socioeconomic Status', *British Journal of Obstetrics and Gynaecology* 90(4), 1983, 291–6.

40 Phyllis B. Eveleth and James M. Tanner, *Worldwide Variation in Growth and Development*, Cambridge University Press, Cambridge, 1990; *World View 1984*, ed. Pete Ayrton, Pluto, London 1983.

41 J. P. Doornbos, et al., 'Differential Birthweights and the Clinical Relevance of Birthweight Standard in a Multiethnic Society', *International Journal of Gynaecology and Obstetrics* 34(4), 1991, 319–24.

42 I. Dawson et al., 'Birthweight by Gestational Age and its Effect on Perinatal Mortality in White and in Punjabi Births: Experience at a District General Hospital in W. London 1967–1975', *British Journal of Obstetrics and Gynaecology* 89(11), 1982, 896–9; S. K. Jivani, 'Asian Neonatal Mortality in Blackburn', *Archives of Diseases of Childhood* 61(5), 1986, 510–2.

43 Marshall H. Klaus, Avroy A. Fanaroff, *Care of the High-Risk Neonate*, W. B. Saunders, Philadelphia, 1986.

44 E. Alberman, 'Are Our Babies Becoming Bigger?', *Journal of the Royal Society of Medicine* 84(5), 1991, 257–60; National Perinatal Epidemiology Unit, *Developing a National Monitoring System to Assess Child Health. Stage 1: Feasibility Study in the Oxford Region, Preliminary Report*, Oxford, 1994.

45 I. Bernasovský et al., 'Body Characteristics of New-born Roms (Gypsies) from Czechoslovakia', *Homo* 30(3), 1979, 151–153.

46 I. Bernasovský, K. Bernasovská, T. Hudáková, 'Some Body Characteristics of Rom (Gypsy) Newborns and their Mothers', *Anthropologie* XIX(3), 1981, 263–268.

47 Lech Mróz, 'Gypsies and the Law', *Ethnologia Polonia* 3 (English language version), 1977, 175–83.

48 I. Bernasovský et al., 'Etudes de quelques parametres biologiques sur les enfants

tsiganes en Slovaquie orientale', presented to V Congreso Español de Antropologia Biologica, Universidad de León, 1987.

49 Eveleth and Tanner, op. cit.

50 Ibid.

51 *Struggling for Identity: Czechoslovakia's Endangered Gypsies*, Helsinki Watch, Human Rights Watch, New York, 1992.

52 Powell, op. cit.

53 *Prague Post*, 12–18 January 1994.

54 Reuter, 23 September 1994.

55 Powell, op. cit.

56 Ian Hancock, pers. comm.

57 *Prague Post*, 2–8 December 1992.

58 Jan Kalanin et al., 'Gypsy Phenylketonuria: A Point Mutation of the Phenylalanine Hydroxylase Gene in Gypsy Families from Slovakia', *American Journal of Medical Genetics* 49(2), 1994, 235–239.

59 Levene et al., op. cit.

60 Ian Hancock, 'Anti-Gypsyism in the New Europe', *Roma* 38/39, 1993, 5–29.

61 Colin Renfrew, *Archaeology and Language*, Jonathan Cape, London, 1987.

62 Roger Pearson, 'Some Comments on Lynn's Thesis by an Anthropologist', *Mankind Quarterly* 32 (1–2), 1991, 175–188.

63 Jean Haudry, *Les Indo-Européens*, Presses Universitaires de France, Paris, 1992.

64 The book was linked to a dispute over the PUF's assocation with Pascal Gauchon, the former leader of a radical right party with affinities to the *Nouvelle Droite*. See *Globe Hebdo*, 10–16 February 1993.

65 Jean-François Kahn, *L'Événement du Jeudi*, 3–9 November 1994.

66 James D. Thomas et al., 'Disease, Lifestyle, and Consanguinity in 58 American Gypsies,' *Lancet*, August 15 1987, 377–379.

67 A. Sutherland, 'Gypsies and Health Care', *Western Journal of Medicine* 1992 September 157(3) 276–80; A. Bodner, M. Leininger, 'Transcultural Nursing Care Values, Beliefs, and Practices of American (USA) Gypsies', *Journal of Transcultural Nursing* 4(1), 1992, 7–18.

68 E. Mair Williams in *Genetic and Population Studies in Wales*, eds Peter S. Harper and Eric Sunderland, University of Wales Press, Cardiff, 1986.

69 Ibid.; E. M. Williams, P.S. Harper, 'Genetic Study of Welsh Gypsies', *Journal of Medical Genetics* 14, 1977, 172–6; V. Ferák, D. Siváková, Z. Sieglová, 'Slovenski Cigani (Romovia) – populacia s najvyssim koeficientom inbridingu v Europe', *Bratislavské Lekárske Listy* 87(2), 1987, 1; M. Lescisinova et al., op. cit.

70 J. Kalanin et al., op. cit.

71 Thomas Acton, 'The Divide Between Social and Physical Anthropology: Some

Reflections on the Work of E. Mair Williams and Judith Okely' (unpublished paper).

—

10 THE BIRTH OF A NATION

1 *The Times Guide to the Peoples of Europe*, ed. Felipe Fernández-Armesto, Times Books, London, 1994.

2 Ivan Bernasovský, application for financial support for research project 'Ethnogenesis of Romanies', Košice, n.d.

3 Thomas Acton, *Gypsy Politics and Social Change: the Development of Ethnic Ideology and Pressure Politics among British Gypsies from Victorian Reformism to Romani Nationalism*, Routledge & Kegan Paul, London, 1974.

4 William G. Lockwood in *Nomadic Peoples*, no. 21/22, December 1986, *Peripatetic Peoples*, ed. Joseph C. Berland, Matt T. Salo, Commission on Nomadic Peoples, Montreal.

5 Ian Hancock, 'The Gypsies: Indian World Citizens', in *Global Migrations of Indians: Saga of Adventure, Enterprise, Identity and Integration*, ed. Jagat K. Motwani, GOPIO, New Delhi, 1993; pers. comm.

6 pers. comm.

7 *Guardian*, 3 May 1995.

8 Acton, op. cit.; 'The Divide Between Social and Physical Anthropology: Some Reflections on the Work of E. Mair Williams and Judith Okely' (unpublished paper).

9 Acton, op. cit.

10 *Patterns of Prejudice* 22(3), 1988, 51.

11 Roger Omond, *The Apartheid Handbook*, Penguin, Harmondsworth, Middlesex, 1985.

12 L. Luca Cavalli-Sforza, Paolo Menozzi and Alberto Piazza, *History and Geography of Human Genes*, Princeton University Press, Princeton, New Jersey, 1994.

13 Ian Hancock, 'The Hungarian Student Valyi Istvan and the Indian Connection of Romani', *Roma* 40, January 1994, 13–15.

14 Ivan Bernasovský, *Seroanthropology of Roms (Gypsies)*, P. J. Šafárik University, Košice, 1994.

15 Milena Hübschmannová, Foreword, *Romano Džaniben 1/1994*; discussion with author.

16 Colin Renfrew, *Archaeology and Language*, Jonathan Cape, London, 1987.

17 Donald Kenrick and Grattan Puxon, *The Destiny of Europe's Gypsies,* Sussex University Press/Heinemann, London, 1972.

18 Bernasovský, op. cit.

19 *Cambridge Encyclopedia of Human Evolution*, eds Steve Jones, Robert Martin and

David Pilbeam, Cambridge University Press, Cambridge, 1992; L. Luca Cavalli-Sforza et al., op. cit.

20 I. Bernasovský et al., 'Body Characteristics of New-born Roms (Gypsies) from Czechoslovakia', *Homo* 30(3), 1979, 151–153.

21 Ian Hancock, 1993, op. cit.

22 Milena Hübschmannová, discussion with author.

23 *Independent*, 13 February 1995; *Searchlight*, March 1995.

24 Sarabjit S. Mastana and Surinder S. Papiha, 'Origin of the Romany Gypsies – Genetic Evidence', *Zeitschrift für Morphologie und Anthropologie* 79(1), 1992, 43–51.

25 Ian Hancock, pers. comm.

26 Thomas Acton, 'The Divide Between Social and Physical Anthropology: Some Reflections on the Work of E. Mair Williams and Judith Okely' (unpublished paper).

11 THE SEACOAST OF MACEDONIA

1 Ludwik Hirschfeld [sic] and Hanka Hirschfeld, 'Serological Differences between the Blood of Different Races: The Result of Researches on the Macedonian Front', *Lancet*, ii, 18 October 1919, 675–679.

2 Raul Hilberg, *Perpetrators, Victims, Bystanders: The Jewish Catastrophe 1939–1945*, Lime Tree, London, 1993; Władysław Bartoszewski and Zofia Lewin (eds), *Righteous Among Nations: How Poles Helped the Jews 1939–1945*, Earlscourt, London 1969. The Hirszfelds were saved from deportation by Polish friends; after the war, Ludwik founded the Immunological Institute in Wrocław.

3 'Sir Walter's Journey', producer Tim Haines, *Horizon*, BBC, 1994.

4 P. Rudan et al., 'Population Structure in the Eastern Adriatic: the Influence of Historical Processes, Migration Patterns, Isolation and Ecological Pressures, and their Interaction', in *Isolation, migration and health: 33rd Symposium Volume of the Society for the Study of Human Biology*, eds D. F. Roberts, N. Fujiki and K. Torizuka, Cambridge University Press, Cambridge, 1992.

5 Petar Vlahović et al., 'Bioantropološka Istraživanja Đerdapskog Podunavlja/ Bioanthropological investigations of the Djerdap region (Danube River Valley)', *Etnoantropološki Problemi* book 2, Department of Ethnology, Faculty of Philosophy in Belgrade, 1988.

6 *Independent*, London, 27 August 1993.

7 Ivan Čolović, 'Ceux qui pètent le feu', *Les Temps Modernes*, February 1993.

8 Ivan Čolović, 'Les mythes politiques du nationalisme ethnique', *Transeuropéennes* No. 3, Paris, Spring 1994.

9 Ivan Čolović, 'Rassismus in elf Bildern', *Perspektiven* Nr. 18, December 1993

(originally published in *Republika,* Belgrade, October 1993). The quote is from *Svet,* Novi Sad, 6 September 1993.

10 Ibid.

11 Ibid.; 'Les mythes politiques du nationalisme ethnique', *Transeuropéennes* No. 3, Paris, Spring 1994.

12 *Perspektiven.* Original quote from *Politika,* 9 October 1993.

13 *La Règle du Jeu,* May 1993.

14 Vlahović et al., op. cit.

15 Elazar Barkan, *The Retreat of Scientific Racism: Changing Concepts of Race in Britain and the United States between the World Wars,* Cambridge University Press, Cambridge, 1992.

16 R.N. Bradley, *Racial Origins of English Character,* George Allen & Unwin, London, 1926.

17 W.W. Howells, in *Cambridge Encyclopedia of Human Evolution,* S. Jones et al. (eds), Cambridge University Press, Cambridge, 1992.

18 Stanley M. Garn, comment on 'Issues in the Study of Race', *Current Anthropology* 3(1), February 1962, 27.

19 Nancy Stepan, *The Idea of Race in Science: Great Britain 1800–1960,* Macmillan, London, 1982.

20 Colin Renfrew, 'Archaeological, Genetic and Linguistic Diversity', *Man/Journal of the Royal Anthropological Institute* 27(3), 1992, 445–478.

21 *Sunday Telegraph,* 27 March 1994.

22 Leon Poliakov, *The Aryan Myth,* Chatto/Heinemann for Sussex University Press, London, 1974.

23 *Newsweek,* 26 July 1993.

24 Hans Magnus Enzensberger, *Europe, Europe: Forays into a Continent,* Hutchinson Radius, London, 1989.

25 Bradley, op. cit.

26 *Independent,* 12 September 1994.

27 A. Piazza et al. 'A Genetic History of Italy', *Annals of Human Genetics* 52, 1988, 203–213.

28 *Guardian,* 1 December 1993.

29 *Searchlight,* September 1994.

30 'Explanation of the Aims and Principles of the Northern League, a North-European Cultural Society', n.d.

12 THROUGH THE BOTTLENECK

1 *Genetic Variation in Britain,* eds D.F. Roberts and E. Sunderland, Taylor & Francis, London, 1973.

2 L.L. Cavalli-Sforza, A.C. Wilson, C.R. Cantor, R.M. Cook-Deegan, M.-C.

King, 'Call for a Worldwide Survey of Human Genetic Diversity: A Vanishing Opportunity for the Human Genome Project', *Genomics* 11(2), 1991, 490–491.

3 *Science* 257, 28 August 1992, 1204–1205; *New Scientist*, 29 May 1993.

4 *New Scientist*, 7 August 1993, 49.

5 Alan Swedlund, 'Is There an Echo in Here? Historical Reflections on the Human Genome Diversity Project', unpublished paper, 1993.

6 F.C. Boyd, *Genetics and the Races of Man*, Little, Brown, Boston, 1950.

7 Walter Bodmer and Robin McKie, *The Book of Man: The Quest to Discover Our Genetic Heritage*, Little, Brown, London, 1994.

8 World Council of Indigenous Peoples, Resolution WCIP/VII/GUA/1993/2-A.

9 Statement Presented by the Indigenous Peoples to the High-Level Meeting of the First Substantive Session of the Commission on Sustainable Development, 24 June 1993.

10 The Human Genome Diversity (HGD) Project, Summary Document, Human Genome Organisation, London, 1994.

11 Letter, 6 July 1993; *Sunday Express* (Dominica), 25 July 1993.

12 *Tropical Star*, 28 July 1993.

13 *New Chronicle*, 23 July 1993.

14 Central Australian Aboriginal Congress, Inc., press release, 'The "Vampire" Project', 24 January 1994.

15 RAFI Communiqué, 'The Patenting of Human Genetic Material', January/February 1994.

16 'File on Four', BBC Radio 4, transmitted 3 December 1994.

17 Report on 'Big Science', produced by Wall to Wall for BBC, transmitted 13 January 1994.

18 Pat Roy Mooney, letters to Jean Doble (Stanford University), 2 June 1993, 6 April 1993.

19 RAFI Communiqué, 'Patents, Indigenous Peoples, and Human Genetic Diversity', May 1993.

20 Aboriginal Health Research Ethics Committee, Aboriginal Health Council of SA, letter 'To Whom It May Concern: Human Genome Diversity', 16 July 1994.

21 Interview with author.

22 Pat Roy Mooney, letter to Dr Henry Greely, 30 June 1993.

23 The Human Genome Diversity (HGD) Project, Summary Document, Human Genome Organisation, London, 1994.

24 Steven Shackley, 'Relics, Rights and Regulations', *Scientific American,* March 1995.

25 'Bones of Contention', written and produced by Danielle Peck and Alex Seaborne, *Horizon* BBC, 1995.

26 *Observer*, 16 April 1995.

27 Richard Cooper and Charles Rotimi, 'Hypertension in Populations of West African Origin: Is There a Genetic Predisposition?', *Journal of Hypertension* 12, 1994, 215–227.

28 Thomas W. Wilson and Clarence E. Grim, 'Biohistory of Slaves and Blood Pressure Differences in Blacks Today: A Hypothesis', *Hypertension* 17 (suppl I), 1991, I-122-I-128.

29 Michael Billig, *Psychology, Racism and Fascism*, A. F. and R. Publications, Birmingham, 1979.

30 Philip D. Curtin, 'The Slavery Hypothesis for Hypertension among African Americans: The Historical Evidence', *American Journal of Public Health* 82(12), 1992, 1681–1686.

31 Fatimah Linda Collier Jackson, 'An Evolutionary Perspective on Salt, Hypertension, and Human Genetic Variability', *Hypertension* 17 (suppl I), 1991, I-129-I-132.

32 Fatimah Linda Collier Jackson, 'The Influence of Dietary Cyanogenic Glycosides from Cassava on Human Metabolic Biology and Microevolution', in *Tropical Forests, Plants and Food: Biocultural Interactions and Applications to Development*, eds C. M. Hladik et al., Unesco/Parthenon, Paris, 1993.

33 Edward O. Wilson, *The Diversity of Life*, Allen Lane, London, 1992.

13 ONE RACE, ONE SCIENCE

1 John Horgan, 'Eugenics Revisited', *Scientific American*, June 1993.

2 Ibid.; C. C. H. Cook and H. M. D. Gurling, 'The Genetic Aspects of Alcoholism and Substance Abuse: A Review', in *The Nature of Drug Dependence*, eds Griffith Edwards and Malcolm Lader, Oxford University Press, Oxford, 1990.

3 Ernest P. Noble et al., 'Allelic Association of the D_2 Dopamine Receptor Gene with Cocaine Dependence', *Drug and Alcohol Dependence* 33, 1993, 271–285.

4 Stevens S. Smith et al., 'The D_2 Dopamine Receptor Taq I B1 RFLP Appears More Frequently in Polysubstance Abusers', *Archives of General Psychiatry* 49, 1992, 723–727 (Reference to unpublished data of B. F. O'Hara, 1991).

5 Horgan, op. cit.

6 *Understanding and Preventing Violence*, National Academy of Sciences and National Research Council, 1992; cited in 'The Resurgence of Genetic Determinism', Dick David, Chicago Coalition Against the Violence Initiative (Compuserve electronic bulletin board message, 1994).

7 Benno Müller-Hill, 'The Shadow of Genetic Injustice', *Nature* 362, 8 April 1993, 491–492.

8 Horgan, op. cit.

Notes and References

9 Benno Müller-Hill, *Murderous Science: Elimination by Scientific Selection of Jews, Gypsies and Others, Germany 1933–1945,* Oxford University Press, Oxford, 1988.

10 Kenan Malik, *The Meaning of Race: Race, History and Culture in Western Society,* Macmillan, London, 1996.

11 Edward O. Wilson, *The Diversity of Life,* Allen Lane, London, 1992.

12 *New Yorker,* 28 November 1994.

13 See G. Ellis Cashmore (ed.), *Dictionary of Race and Ethnic Relations,* Routledge, London, 1988.

14 For a discussion of the 'gay gene' question, see Daniel J. Kevles, 'The X Factor' (book review), *New Yorker,* 3 April 1995, 85–90.

15 Sandra Scarr, 'Three Cheers for Behavior Genetics: Winning the War and Losing Our Identity', *Behavior Genetics* 17(3), 1987, 219–228.

16 Karl Marx, from *The Paris Manuscripts,* 1844, quoted by Pierre L. van den Berghe in *The Ethnic Phenomenon,* Elsevier, New York, 1981.

17 'The Resurgence of Genetic Determinism', Dick David, Chicago Coalition Against the Violence Initiative (Compuserve electronic bulletin board message, 1994). See also Jonathan Marks, 'Black, White, Other', *Natural History* December 1994.

Index

Index

Index

Also available in Vintage

Nicholas Humphrey

A HISTORY OF THE MIND

Winner of the 1993 British Psychological Society Award

The link between our physical being and our consciousness of what it is like to be ourselves – the mind-body problem – has baffled generations of philosophers and remains the greatest challenge to contemporary science. Nicholas Humphrey tells the story of how the mind-body link has been forged by evolution and argues for a new theory of how feelings enter consciousness.

'Exceptionally readable, and packed with fascinating psychological information and ingenious speculation...Humphrey writes with an unusual combination of verve, lucidity and charm'

Michael Lockwood, *Guardian*

'An eloquent and persuasive theory about how the water of the physical brain is turned into the wine of consciousness. No other theoretical psychologist is so accessibly clear, and at the same time so provocatively philosophical'

Lorna Sage, *Observer*

VINTAGE

A SELECTED LIST OF SCIENCE TITLES
AVAILABLE IN VINTAGE

☐ THEORIES OF EVERYTHING	John Barrow	£6.99
☐ THE RISE AND FALL OF THE THIRD CHIMPANZEE	Jared Diamond	£7.99
☐ COMING OF AGE IN THE MILKY WAY	Timothy Ferris	£8.99
☐ 'SURELY YOU'RE JOKING, MR FEYNMAN'	Richard P. Feynman	£7.99
☐ ALAN TURING – THE ENIGMA	Andrew Hodges	£8.99
☐ SOUL SEARCHING	Nicholas Humphrey	£7.99
☐ A HISTORY OF THE MIND	Nicholas Humphrey	£7.99
☐ THE INNER EYE	Nicholas Humphrey	£4.99
☐ THE FIFTH ESSENCE	Lawrence Krauss	£7.99
☐ FEAR OF PHYSICS	Lawrence Krauss	£7.99
☐ SHADOWS OF THE MIND	Roger Penrose	£7.99
☐ THE EMPEROR'S NEW MIND	Roger Penrose	£9.99
☐ DREAMS OF A FINAL THEORY	Steven Weinberg	£6.99
☐ THE BEAK OF THE FINCH	Jonathan Weiner	£8.99

- All Vintage books are available through mail order or from your local bookshop.

- Please send cheque/eurocheque/postal order (sterling only), Access, Visa or Mastercard:

☐☐☐☐☐☐☐☐☐☐☐☐☐☐☐☐

Expiry Date: _____ Signature: _____

Please allow 75 pence per book for post and packing U.K.
Overseas customers please allow £1.00 per copy for post and packing.

ALL ORDERS TO:

Vintage Books, Book Service by Post, P.O.Box 29, Douglas, Isle of Man, IM99 1BQ.
Tel: 01624 675137 • Fax: 01624 670923

NAME: _____

ADDRESS: _____

_____

_____

Please allow 28 days for delivery. Please tick box if you do not
wish to receive any additional information ☐

Prices and availability subject to change without notice.